AMERICA'S LAB REPORT

Investigations in High School Science

Committee on High School Laboratories: Role and Vision

Susan R. Singer, Margaret L. Hilton, and
Heidi A. Schweingruber, Editors

Board on Science Education

Center for Education

Division of Behavioral and Social Sciences and Education

NATIONAL RESEARCH COUNCIL
OF THE NATIONAL ACADEMIES

THE NATIONAL ACADEMIES PRESS
Washington, D.C.
www.nap.edu

THE NATIONAL ACADEMIES PRESS • 500 Fifth Street, N.W. • Washington, D.C. 20001

NOTICE: The project that is the subject of this report was approved by the Governing Board of the National Research Council, whose members are drawn from the councils of the National Academy of Sciences, the National Academy of Engineering, and the Institute of Medicine. The members of the committee responsible for the report were chosen for their special competences and with regard for appropriate balance.

This study was supported by Grant No. ESI-0102582 between the National Academy of Sciences and the National Science Foundation. Any opinions, findings, conclusions, or recommendations expressed in this publication are those of the author(s) and do not necessarily reflect the views of the organizations or agencies that provided support for the project.

Library of Congress Cataloging-in-Publication Data

America's lab report : investigations in high school science / Committee on High School Science Laboratories--Role and Vision, Board on Science Education, Center for Education, Division of Behavioral and Social Sciences and Education; Susan R. Singer, Margaret L. Hilton, and Heidi A. Schweingruber, editors.
   p. cm.
  Includes bibliographical references and index.
  ISBN 0-309-09671-5 —ISBN 0-309-55100-5 (pdf)
  1. Science—Study and teaching (Secondary)—United States. 2. Education, Secondary—Curricula—United States. 3. Laboratories—Curricula—United States.   I. Singer, Susan R. II. Hilton, Margaret L. III. Schweingruber, Heidi A. IV. National Research Council (U.S.). Committee on High School Science Laboratories: Role and Vision.
  Q183.3.A1A44 2006
  507'.1273--dc22

                                                                                    2005026110

Additional copies of this report are available from the National Academies Press, 500 Fifth Street, N.W., Lockbox 285, Washington, DC 20055; (800) 624-6242 or (202) 334-3313 (in the Washington metropolitan area); Internet, http://www.nap.edu

Copyright 2006 by the National Academy of Sciences. All rights reserved.
Printed in the United States of America.

Suggested citation: National Research Council. (2006). *America's Lab Report: Investigations in High School Science.* Committee on High School Science Laboratories: Role and Vision, S.R. Singer, M.L. Hilton, and H.A. Schweingruber, Editors. Board on Science Education, Center for Education. Division of Behavioral and Social Sciences and Education. Washington, DC: The National Academies Press.

THE NATIONAL ACADEMIES
Advisers to the Nation on Science, Engineering, and Medicine

The **National Academy of Sciences** is a private, nonprofit, self-perpetuating society of distinguished scholars engaged in scientific and engineering research, dedicated to the furtherance of science and technology and to their use for the general welfare. Upon the authority of the charter granted to it by the Congress in 1863, the Academy has a mandate that requires it to advise the federal government on scientific and technical matters. Dr. Ralph J. Cicerone is president of the National Academy of Sciences.

The **National Academy of Engineering** was established in 1964, under the charter of the National Academy of Sciences, as a parallel organization of outstanding engineers. It is autonomous in its administration and in the selection of its members, sharing with the National Academy of Sciences the responsibility for advising the federal government. The National Academy of Engineering also sponsors engineering programs aimed at meeting national needs, encourages education and research, and recognizes the superior achievements of engineers. Dr. Wm. A. Wulf is president of the National Academy of Engineering.

The **Institute of Medicine** was established in 1970 by the National Academy of Sciences to secure the services of eminent members of appropriate professions in the examination of policy matters pertaining to the health of the public. The Institute acts under the responsibility given to the National Academy of Sciences by its congressional charter to be an adviser to the federal government and, upon its own initiative, to identify issues of medical care, research, and education. Dr. Harvey V. Fineberg is president of the Institute of Medicine.

The **National Research Council** was organized by the National Academy of Sciences in 1916 to associate the broad community of science and technology with the Academy's purposes of furthering knowledge and advising the federal government. Functioning in accordance with general policies determined by the Academy, the Council has become the principal operating agency of both the National Academy of Sciences and the National Academy of Engineering in providing services to the government, the public, and the scientific and engineering communities. The Council is administered jointly by both Academies and the Institute of Medicine. Dr. Ralph J. Cicerone and Dr. Wm. A. Wulf are chair and vice chair, respectively, of the National Research Council

www.national-academies.org

COMMITTEE ON HIGH SCHOOL SCIENCE LABORATORIES: ROLE AND VISION

SUSAN R. SINGER (*Chair*), Department of Biology, Carleton College
HUBERT M. DYASI, School of Education, City College of the City University of New York
ARTHUR EISENKRAFT, Center of Science and Mathematics in Context, University of Massachusetts, Boston
PAMELA J. HINES, *Science*/American Association for the Advancement of Science, Washington, DC
MICHAEL LACH, Chicago Public Schools
DAVID PAUL LICATA, Pacifica High School, Garden Grove, CA
NANCY PELAEZ, Department of Biological Science, California State University, Fullerton
WILLIAM A. SANDOVAL, Graduate School of Education and Information Studies, University of California, Los Angeles
JAMES P. SPILLANE, Institute for Policy Research, Northwestern University
CARL E. WIEMAN, Department of Physics, University of Colorado, Boulder

MARGARET L. HILTON, *Study Director*
HEIDI A. SCHWEINGRUBER, *Program Officer*
JEAN MOON, *Director, Board on Science Education*
MARY ANN KASPER, *Senior Program Assistant*

BOARD ON SCIENCE EDUCATION

CARL E. WEIMAN *(Chair)*, Department of Physics and JILA, University of Colorado, Boulder
PHILIP BELL, Cognitive Studies in Education, University of Washington, Seattle
KATHLEEN COMFORT, WestEd, San Francisco
DAVID T. CONLEY, Center for Educational Policy Research, University of Oregon, Eugene
BARBARA L. GONZALEZ, Department of Chemistry and Biochemistry, California State University, Fullerton
LINDA D. GREGG, TERC, Cambridge, MA
JENIFER V. HELMS, Education Consultant, Denver, CO
JOHN R. JUNGCK, Biology Department, Beloit College
ISHRAT M. KHAN, Department of Chemistry, Clark Atlanta University
OKHEE LEE, Department of Teaching and Learning, University of Miami, Coral Gables
SHARON LONG, Department of Biological Sciences, Stanford University
RICHARD A. McCRAY, Department of Astrophysical and Planetary Sciences and JILA, University of Colorado, Boulder
LILLIAN C. McDERMOTT, Department of Physics, University of Washington, Seattle
MARY "MARGO" MURPHY, Georges Valley High School, Thomaston, ME
CARLO PARRAVANO, Merck Institute for Science Education, Merck & Co., Inc., Rahway, NJ
MARY JANE V. SCHOTT, The Charles A. Dana Center, University of Texas, Austin
SUSAN R. SINGER, Department of Biology, Carleton College
CARY SNEIDER, Boston Museum of Science

JEAN MOON, *Director*
HEIDI A. SCHWEINGRUBER, *Program Officer*
ANDREW SHOUSE, *Program Officer*
LaSHAWN SIDBURY, *Senior Program Assistant*

Foreword

It will soon be 25 years since Terrell H. Bell, Secretary of Education in the Reagan administration, commissioned a task force to examine the state of education in the United States. The work of this commission resulted in the 1983 report *A Nation at Risk: An Imperative for Educational Reform*, which detailed what was then a shocking report card on American education. The report became not only a rallying cry for an improved and equitable system of education but also an early framework for education reform. Regarding high school science education, *A Nation at Risk* made the following recommendation:

> The teaching of science in high school should provide graduates with an introduction to: (a) the concepts, laws, and processes of the physical and biological sciences; (b) the methods of scientific inquiry and reasoning; (c) the application of scientific knowledge to everyday life; and (d) the social and environmental implications of scientific and technological development. Science courses must be revised and updated for both the college-bound and those not intending to go to college. (p. 25)

In the science education community, we continue to be challenged by the goals for science education set out in *A Nation at Risk*. The call for students to be familiar with the methods of science inquiry and reasoning and to understand the concepts and processes of the sciences remains a visible, but largely unmet, national educational goal. Indeed, this book describes what we know and do not know about the potential of laboratories to serve as effective science learning environments. The book defines such environments as places in which students can practice scientific inquiry and reasoning, come to understand different kinds of knowledge claims that scientists make, and build their knowledge of science content.

Since *A Nation at Risk* was released, the remarkable advances in science and technology have produced even greater public concern over the quality of science education. One has only to think about the human genome project. Completed in April 2003, it provides the complete genetic blueprint for humans. It is hard to comprehend the long-term effects of this kind of scientific advancement. In educational terms, however, such discoveries raise local, state, and national expectations for science education. Today a majority of policy makers, scientists, educators, and parents agree that high school graduates must have a sophisticated grasp and appreciation of science and technology to participate fully in the work place, to understand their everyday decisions on matter ranging from health to energy resources to climate, and to participate as informed citizens in the civic realm.

Interest in science education is shared around the world, whether the country is industrialized or developing. It seems universally understood that effective science education is a critical component for advancing scientific and societal development. In the United States, laboratories have been a part of science education since the late 1800's. Though educational goals for labs have shifted over time as have instructional materials and laboratory equipment, their presence as part of high school science has been consistent. Given the long history of laboratories in school science, the absence of consistent and well-grounded research on high school labs is troubling. *America's Lab Report* begins to fill this important void.

America's Lab Report is the first consensus study to be completed under the guidance of the Board on Science Education. On behalf of the board, we want to thank the ten experts who served on the study committee. Each study committee member brought a wealth of knowledge about the nature and enterprise of science, the teaching and learning of science, and the institutions of schools and schooling to their deliberations. It was a very thoughtful group of committee members who took their charge very seriously.

Chair of the study committee, Susan Singer, warrants special acknowledgment. Being chair for a National Research Council study is a time-consuming commitment and one that invites patience. Susan's persistence and insight into the process engendered a great deal of respect. The entire committee process was helped by the skillful work of Margaret Hilton, study director, and program officer Heidi Schweingruber. Each brought a unique set of talents to their work for which I am very grateful.

Finally, on behalf of the Board on Science Education, we want to thank the National Science Foundation staff for their initial conversations on this very challenging topic, their turning to the board to undertake this work, and recognition of the board as the right oversight group, and their support of this study.

America's Lab Report: Investigations in High School Science is born of hours of sustained examination of a broad body of evidence by a diverse and uniquely qualified group of experts. The result is a previously unavailable synthesis of research that supports a compelling discussion of the evolving role of laboratories in advancing the goals of science education. Our hope for the report is that, in the spirit of ***A Nation at Risk***, it will catalyze informed debate about laboratories and school science that leads to improvement of science education for our nation's high school students.

Carl E. Wieman
Chair
Board on Science Education

Jean Moon
Director
Board on Science Education

Acknowledgments

The committee and staff thank the many individuals and organizations without whom this study could not have been completed.

First, we acknowledge the support of the National Science Foundation (NSF). We particularly thank NSF program officer Janice Earle, who consistently supported and encouraged the study committee and staff during the past year and a half. We are also grateful to James Lightbourne, who organized discussions among the NSF staff, which led to the request for the study.

Individually and collectively, members of the committee benefited from discussions and presentations by the many individuals who participated in our three fact-finding meetings. At the first meeting, the following individuals informed the committee about key issues affecting teaching and learning in high school science laboratories: David Hammer, associate professor of physics and of curriculum and instruction, University of Maryland; Sean Smith, senior research associate, Horizon Research, Inc.; Gerald F. Wheeler, executive director, National Science Teachers Association; Warren W. Hein, associate executive director, American Association of Physics Teachers; Angela Powers, senior education associate, teacher training, American Chemical Society; Michael J. Smith, former education director, American Geological Institute. We also thank Janet Carlson Powell, associate director, Biological Sciences Curriculum Study; Robert Tinker, president, The Concord Consortium; Jo Ellen Roseman, director, Project 2061; and George De Boer, deputy director, Project 2061, American Association for the Advancement of Science, for briefing the committee on the role of science curriculum materials and technology in high school science laboratory activities.

At its second meeting, the committee learned about a variety of factors influencing high school science laboratories, ranging from the nature of science to technology to state science assessments. We are grateful to each of the presenters, including: Jane Maienschein, professor and director of the Center for Biology and Society, Arizona State University; Robin Millar, professor of science education, University of York; Arthur Lidsky, president, Dober, Lidsky, Craig and Associates; Adam Gamoran, professor of sociology, University of Wisconsin; Marcia Linn, professor of development and cognition, University of California, Berkeley; Kefyn Catley, assistant professor of science education, Vanderbilt University; Mark Windschitl, associate professor, College of Education, University of Washington; Audrey Champagne, professor, Department of Educational Theory and Practice and Department of Chemistry, State University of New York at Albany; Thomas Shiland, science department chair, Saratoga Springs High School, NY; and Arthur Halbrook, senior project associate, Council of Chief State School Officers. We also thank the individuals who participated in panels addressing how financial and resource constraints and school organization influence laboratory teaching and learning. The panelists include: Daniel Gohl, principal, McKinley Technical High School, Washington, DC Public Schools; Shelley Lee, science education consultant, Wisconsin Department of Public Instruction; Lynda Beck, former assistant head of school, Phillips Exeter Academy; and Kim Lee, science curriculum supervisor, Montgomery County Public Schools, VA.

We thank the following individuals who shared their expertise on student science learning with the committee at its final fact-finding meeting: Philip Bell, associate professor, College of Education, University of Washington; Richard Duschl, professor of science education, Rutgers University; Norman Lederman, professor of mathematics and science education, Illinois Institute of Technology; Okhee Lee, professor, School of Education, University of Miami; Sharon Lynch, professor of secondary education, George Washington University; Kenneth Tobin, professor of urban education, Graduate Center of City College, NY; Samuel Stringfield, principal research scientist, Johns Hopkins University Center for the Social Organization of Schools; Ellyn Daugherty, lead teacher, San Mateo High School Biotechnology Careers Pathway Program; Elaine Johnson, director, Bio-Link, City College of San Francisco; Robert Tai, assistant professor of science education, University of Virginia. At its last open session, the committee talked with a panel of master science teachers to learn about their approaches to laboratory teaching. We are grateful to each member of the panel, including Nina Hike-Teague, Curie High School, Chicago; Gertrude Kerr, Howard High School, Howard County, MD; Margot Murphy, George's Valley High School, ME; Phil Sumida, Maine West High School, Des Plaines, IL; and Robert Willis, Ballou High School, Washington, DC.

Many individuals at the National Research Council (NRC) assisted the committee. The study would not have been possible without the efforts of Jean Moon, who quickly wrote the initial proposal in response to NSF's request. Patricia Morison offered valuable suggestions at each committee meeting and during the review process, as well as providing helpful comments on several drafts of the report. Eugenia Grohman helped to focus the final committee meeting on the key messages and conclusions emerging from the study. We thank Kirsten Sampson Snyder, who shepherded the report through the NRC review process, Christine McShane, who edited the draft report, and Yvonne Wise for processing the report through final production. At an early stage in the study, Barbara Schulz invited us to a meeting with the Teacher Advisory Council, which provided practical insights into high school science laboratories. Brenda Buchbinder managed the finances of the project, and Viola Horek provided important organizational and administrative assistance. We are grateful to LaShawn Sidbury who arranged logistics for the first committee meeting. Finally, we would like to thank Mary Ann Kasper for her able assistance in supporting the committee at every stage in its deliberations and in preparing numerous drafts and revisions of the report.

This report has been reviewed in draft form by individuals chosen for their diverse perspectives and technical expertise, in accordance with procedures approved by the Report Review Committee of the NRC. The purpose of this independent review is to provide candid and critical comments that will assist the institution in making its published report as sound as possible and to ensure that the report meets institutional standards for objectivity, evidence, and responsiveness to the charge. The review comments and draft manuscript remain confidential to protect the integrity of the deliberative process.

We thank the following individuals for their review of this report: Brian Drayton, Science Education, TERC, Cambridge MA; Kevin Dunbar, Department of Education, Dartmouth College; James W. Guthrie, Department of Leadership, Policy, and Organizations, Vanderbilt University; David G. Haase, Physics and The Science House, North Carolina State University; Thomas E. Keller, Science Education, Maine Department of Education, Augusta, ME; Vincent N. Lunetta, Science Education, and Science, Technology, Society (emeritus), Pennsylvania State University; Arlene A. Russell, School of Education and Chemistry and Biochemistry, University of California, Los Angeles; James H. Stewart, Center for Biology, University of Wisconsin–Madison; Phil Sumida, Science Department, Maine West High School, Des Plaines, IL; and Ellen Weaver, Department of Biology (emeritus), San Jose State University.

Although the reviewers listed above provided many constructive comments and suggestions, they were not asked to endorse the conclusions and recommendations nor did they see the final draft of the report before its release. The review of this report was overseen by Michael E. Martinez, Department of Education, University of California, Irvine, and May Berenbaum, Department of Entomology, University of Illinois. Appointed by the NRC, they were responsible for making certain that an independent examination of this report was carried out in accordance with institutional procedures and that all review comments were carefully considered. Responsibility for the final content of this report rests entirely with the authoring committee and the institution.

Contents

Executive Summary   1

1 Introduction, History, and Definition of Laboratories   13

2 The Education Context   42

3 Laboratory Experiences and Student Learning   75

4 Current Laboratory Experiences   116

5 Teacher and School Readiness for Laboratory Experiences   138

6 Facilities, Equipment, and Safety   168

7 Laboratory Experiences for the 21st Century   193

Appendixes

A Agendas of Fact-Finding Meetings   205

B Biographical Sketches of Committee Members and Staff   215

AMERICA'S LAB REPORT

Executive Summary

Most people in this country lack the basic understanding of science that they need to make informed decisions about the many scientific issues affecting their lives. Neither this basic understanding—often referred to as scientific literacy—nor an appreciation for how science has shaped the society and culture is being cultivated during the high school years. For example, over the 30 years between 1969 and 1999, high school students' scores on the science portion of the National Assessment of Educational Progress (NAEP, the "nation's report card") remained stagnant. In addition, high school students' performance on a different NAEP national science assessment, first administered in 1996, was weaker four years later in 2000. Yet policy makers, scientists, and educators agree that high school graduates today, more than ever, need a basic understanding of science and technology in order to function effectively in an increasingly complex, technological society. Increasing this understanding will require major reforms in science education, including reforms in the laboratories that constitute a significant portion of the high school science curriculum.

Since the late 19th century, high school students in the United States have carried out laboratory investigations as part of their science classes. Educators and policy makers have periodically debated the value of laboratories in helping students understand science, but little research has been done to inform those debates or to guide the design of laboratory education. Today, on average, students enrolled in science classes spend about one class period per week in such laboratory investigations as observing and comparing different cell types under a microscope in biology class or adding a solution of known acidity to a solution of unknown alkalinity in chemistry class. To assess how these and similar laboratory activities may contribute to science learning, the National Science Foundation requested the National Research Council to examine the current status of science laboratories and develop a vision for their future role in high school science education.

DEFINITION AND GOALS OF HIGH SCHOOL SCIENCE LABORATORIES

Questions about the value of high school science laboratories stem in part from a lack of clarity about what exactly constitutes a "laboratory" and what its science learning goals might be. For example, "laboratory" may refer to a room equipped with benches and student workstations, or it may refer to various types of indoor or outdoor science activities. Today and in the past, educators, policy makers, and researchers have not agreed on a common definition of "laboratory."

This lack of clarity about the definition and goals of laboratories has slowed research on their outcomes. In addition, mechanisms for sharing the results of the research that is available—both within the research community and with the larger education community—are so weak that progress toward more effective laboratory learning experiences is impeded.

> *Conclusion 1: Researchers and educators do not agree on how to define high school science laboratories or on their purposes, hampering the accumulation of evidence that might guide improvements in laboratory education. Gaps in the research and in capturing the knowledge of expert science teachers make it difficult to reach precise conclusions on the best approaches to laboratory teaching and learning.*

Rapid developments in science, technology, and cognitive research have made the traditional definition of science laboratories—only as rooms where students use special equipment to carry out well-defined procedures—obsolete. Rather, the committee gathered information on a wide variety of approaches to laboratory education, arriving at the term "laboratory experiences" to describe teaching and learning that may take place in a laboratory room or in other settings.

While the committee found that many laboratory experiences involve students in carrying out carefully specified procedures to verify established scientific knowledge, we also learned of laboratory experiences that engaged students in formulating questions, designing investigations, and creating and revising explanatory models. Participating in a range of laboratory experiences holds potential to enhance students' understanding of the dynamic relationships between empirical research and the scientific theories and concepts that both result from research and lead to further research questions.

Committee Definition of Laboratory Experiences

To frame the scope of the study while also reflecting the variety of laboratory experiences, the committee defined laboratory experiences as follows:

Laboratory experiences provide opportunities for students to interact directly with the material world (or with data drawn from the material world), using the tools, data collection techniques, models, and theories of science.

This definition includes student interaction with astronomical databases, genome databases, databases of climatic events over long time periods, and other large data sets derived directly from the material world. It does not include student manipulation or analysis of data created by a teacher to simulate direct interaction with the material world. For example, if a physics teacher presented students with a constructed data set on the weight and required pulling force for boxes pulled across desks with different surfaces and asked them to analyze these data, the students' problem-solving activity would not constitute a laboratory experience in the committee's definition.

In the committee's view, science education includes learning about the methods and processes of scientific research (science process) and the knowledge derived through this process (science content). Science process centers on direct interactions with the natural world aimed at explaining natural phenomena. Science education would not be about science if it did not include opportunities for students to learn about both the process and the content of science. Laboratory experiences, in the committee's definition, can potentially provide one such opportunity.

Goals of Laboratory Experiences

In our review of the literature, the committee identified a number of science learning goals that have been attributed to laboratory experiences, including:

- enhancing mastery of subject matter;
- developing scientific reasoning;
- understanding the complexity and ambiguity of empirical work;
- developing practical skills;
- understanding the nature of science;
- cultivating interest in science and interest in learning science; and
- developing teamwork abilities.

Helping all high school students achieve these science learning goals is critical to improving national scientific literacy and preparing the next generation of scientists and engineers.

Although no single laboratory experience is likely to achieve all of these learning goals, different types of laboratory experiences may be designed to achieve one or more goals. For example, the committee studied a sequence of laboratory experiences included in a larger unit of instruction. Students predicted the temperatures of everyday objects, tested their predictions using temperature-sensitive probes connected to computers, and developed and revised scientific explanations for their results. Students participating in the laboratory experiences and other learning activities progressed toward two goals. They increased their mastery of subject matter (thermodynamics) and their interest in science in comparison to students who participated in the traditional program of science instruction. Some of the science learning goals presented above, particularly understanding the complexity and ambiguity of empirical work, can be attained only through laboratory experiences.

EFFECTIVENESS OF LABORATORY EXPERIENCES

The committee's review of the evidence on attainment of the goals of laboratory experiences reveals a recent shift in research, reflecting some movement in laboratory instruction. Historically, laboratory experiences have been disconnected from the flow of classroom science lessons. Because this approach remains common today, we refer to these separate laboratory experiences as "typical" laboratory experiences. Reflecting this separation, researchers often engaged students in one or two experiments or other science activities and then conducted assessments to determine whether their understanding of the science concept underlying the activity had increased. Some studies compared the outcomes of these separate laboratory experiences with the outcomes of other forms of science instruction, such as lectures or discussions.

Over the past 10 years, a new body of research on the outcomes of laboratory experiences has been developing. Drawing on principles of learning derived from the cognitive sciences, researchers are investigating *how* to sequence science instruction, including laboratory experiences, in order to support students' science learning. We propose the phrase "integrated instructional units" to describe these sequences of instruction. Integrated instructional units connect laboratory experiences with other types of science learning activities, including lectures, reading, and discussion. Students are engaged in framing research questions, making observations, designing and executing experiments, gathering and analyzing data, and constructing scientific arguments and explanations.

Integrated instructional units are designed to increase students' ability to understand and apply science subject matter (often focusing on one important concept or principle) while also improving their scientific reasoning, interest in science, and understanding of the nature of science. Students are encouraged to discuss their existing ideas about the science concept and their emerging ideas during the course of their laboratory experiences, both with their peers and with the teacher. The sequence of laboratory experiences and other forms of instruction is designed to help students develop a more sophisticated understanding of both the science concept under study and the process through which scientific concepts are developed, evaluated, and refined.

The earlier body of research on typical laboratory experiences and the emerging research on integrated instructional units yield different findings about the effectiveness of laboratory experiences in advancing the goals identified by the committee. Research on typical laboratory experiences is methodologically weak and fragmented, making it difficult to draw precise conclusions. The weight of the evidence from research focused on the goals of developing scientific reasoning and cultivating student interest in science shows slight improvements in both after students participated in typical laboratory experiences. Research focused on the goal of student mastery of subject matter indicates that typical laboratory experiences are no more or less effective than other forms of science instruction (such as reading, lectures, or discussion).

A major limitation of the research on integrated instructional units is that most of the units have been used in small numbers of science classrooms. Only a few studies have addressed the challenges of implementing—and studying the effectiveness of—integrated instructional units on a wide scale. The studies conducted to date indicate that these sequences of laboratory experiences and other forms of instruction show greater effectiveness for these same three goals (compared with more traditional forms of science instruction): improving mastery of subject matter, developing scientific reasoning, and cultivating interest in science. Integrated instructional units also appear to be effective in helping diverse groups of students progress toward these three learning goals. Due to a lack of available studies, the committee was unable to draw conclusions about the extent to which either typical laboratory experiences or integrated instructional units might advance the other goals identified at the beginning of this chapter—enhancing understanding of the complexity and ambiguity of empirical work, acquiring practical skills, and developing teamwork skills.

The committee considers the evidence emerging from research on integrated instructional units sufficient to conclude:

Conclusion 2: Four principles of instructional design can help laboratory experiences achieve their intended learning goals if: (1) they are designed with clear learning outcomes in mind, (2) they are thoughtfully sequenced into the flow of classroom science instruction, (3) they are designed to integrate learning of science content with learning about the processes of science, and (4) they incorporate ongoing student reflection and discussion.

CURRENT HIGH SCHOOL LABORATORY EXPERIENCES

Most science students in U.S. high schools today participate in laboratory experiences that are isolated from the flow of classroom science instruction (referred to here as "typical" laboratory experiences). Instead of focusing on clear learning goals, teachers and laboratory manuals often emphasize the procedures to be followed, leaving students uncertain about what they are supposed to learn. Lacking a focus on learning goals related to the subject matter being addressed in the science class, these typical laboratory experiences often fail to integrate student learning about the processes of science with learning about science content. Typical laboratory experiences rarely incorporate ongoing reflection and discussion among the teacher and the students, although there is evidence that reflecting on one's own thinking is essential for students to make meaning out of their laboratory activities. In general, most high school laboratory experiences do not follow the instructional design principles for effectiveness identified by the committee. In addition, most high school students participate in a limited range of laboratory activities that do not help them to fully understand science process.

Several factors contribute to the prevalence of typical laboratory experiences. These include a lack of preparation of—and support for—teachers, disparities in the availability and quality of laboratory facilities and equipment, interpretations of state science standards, and the lack of agreement on definitions and goals of laboratory experiences. Students in schools with higher concentrations of non-Asian minorities spend less time in laboratory instruction than students in other schools, and students in lower level science classes spend less time in laboratory instruction than those enrolled in more advanced science classes. And some students have no access to any type of laboratory experience. Taken together, all of these factors weaken the effectiveness of current laboratory experiences.

Conclusion 3: The quality of current laboratory experiences is poor for most students.

Teacher Preparation for Laboratory Experiences

Teachers play a critical role in leading effective laboratory experiences. By carefully introducing the experiences in ways that are aligned with the learning goals of the science course and leading discussions and answering questions, the teacher can support students in linking their laboratory experiences to underlying science concepts. By selecting laboratory experiences that are clearly related to the ongoing flow of classroom science instruction, the teacher can integrate student learning of both the processes of science and important science content. Yet the undergraduate education of future high school science teachers does not currently prepare them with the pedagogical and science content knowledge required to carry out such teaching strategies. Undergraduate science departments rarely provide future science teachers with laboratory experiences that follow the design principles derived from recent research—integrated into the flow of instruction, focused on clear learning goals, aimed at the learning of science content and science process, with ongoing opportunities for reflection and discussion.

Once on the job, science teachers have few opportunities to improve their laboratory teaching. Professional development opportunities for science teachers are limited in quality, availability, and scope and place little emphasis on laboratory instruction. In addition, few high school teachers have access to curricula that integrate laboratory experiences into the stream of instruction, although such curricula might help them in improving the instructional quality of laboratory experiences. Few high schools support science teachers in improving their laboratory teaching by providing appropriate, ongoing professional development, well-designed science curricula, and adequate laboratory facilities and supplies.

Conclusion 4: Improving high school science teachers' capacity to lead laboratory experiences effectively is critical to advancing the educational goals of these experiences. This would require major changes in undergraduate science education, including providing a range of effective laboratory experiences for future teachers and developing more comprehensive systems of support for teachers.

Laboratory Facilities and School Organization

The capacity of teachers and schools to advance the learning goals of laboratory experiences is affected by laboratory facilities and supplies and the organization of schools.

Direct observation and manipulation of many aspects of the material world require adequate laboratory facilities, including space for teacher demonstrations, student laboratory activities, student discussion, and safe storage space for supplies. Schools with higher concentrations of non-Asian minorities and schools with higher concentrations of poor students are less likely to have adequate laboratory facilities than other schools. In addition to less adequate laboratory space, schools with higher concentrations of poor or minority students and rural schools often have lower budgets for laboratory equipment and supplies than other schools. These disparities in facilities and supplies may contribute to the problem that students in schools with high concentrations of non-Asian minority students spend less time in laboratory instruction than students in other schools.

The ability of schools to address the pressing need for improvements in laboratory teaching is constrained by the way many schools are organized. Often, administrators, teachers, and students become accustomed to routines in class schedules, teachers' schedules, the allocation of space, supplies, and budgets, and teaching approaches. When such routines become rigid, they tend to reinforce existing knowledge and teaching practices, limiting teachers' and administrators' motivation and ability to try out new, more effective approaches to laboratory education. For example, routines in class scheduling and space allocation may limit science teachers' ability or willingness to collaborate with other teachers in shared lesson planning, reflection, and improvement of laboratory lessons. Teachers and administrators who are accustomed to their existing science texts and laboratory manuals may not seek information about new science curricula that effectively integrate laboratory experiences, or they may hesitate to implement such curricula. Rigid school schedules may discourage teachers from adopting new, more effective approaches to laboratory instruction when such approaches require extended classroom time for students and teachers to discuss and reflect on the meaning of laboratory investigations.

Conclusion 5: The organization and structure of most high schools impedes teachers' and administrators' ongoing learning about science instruction and ability to implement quality laboratory experiences.

State Standards and Accountability Systems

Most states have developed science standards to guide instruction and large-scale assessments to measure attainment of those standards. These standards could be used as flexible frameworks to guide schools and teachers in integrating laboratory experiences into the flow of instruction in order to help students master science subject matter while also developing scientific reasoning and advancing other learning goals. However, this rarely happens. Instead, state and local officials and science teachers often see state standards as requiring them to help students master the specific science topics outlined for a grade level or science course. When they view laboratory experiences as isolated events that do not contribute to mastery of topics and science class time is short, laboratory experiences may be limited. For example, research on integrated instructional units has shown that engagement with laboratory experiences and other forms of instruction over periods of 6 to 16 weeks can increase students' mastery of a complex science topic, including the relationships among scientific ideas related to that topic. But teachers who try to "cover" an extensive list of science topics included in state science standards within a school year may have only a few days for each topic, precluding use of such potentially effective instructional units.

The interpretation and implementation of state science standards may also limit attainment of the educational goals of laboratory experiences in other ways. When state standards are seen primarily as lists of science topics to be mastered, they support attainment of only one of the many goals of laboratory experiences—mastery of subject matter. Some state standards call for students to engage in laboratory experiences and to attain other goals of laboratory experiences, such as developing scientific reasoning and understanding the nature of science. However, assessments in these states rarely include items designed to measure student attainment of these goals.

Conclusion 6: State science standards that are interpreted as encouraging the teaching of extensive lists of science topics in a given grade may discourage teachers from spending the time needed for effective laboratory learning.

Conclusion 7: Current large-scale assessments are not designed to accurately measure student attainment of the goals of laboratory experiences. Developing and implementing improved assessments to encourage effective laboratory teaching would require large investments of funds.

WHAT NEXT? RESEARCH, DEVELOPMENT, AND IMPLEMENTATION OF EFFECTIVE LABORATORY EXPERIENCES

Laboratory experiences have the potential to help students attain several important learning goals, including mastery of science subject matter, increased interest in science, and development of scientific reasoning skills. That potential is not being realized today.

The committee does not recommend any specific policies or programs to enhance the effectiveness of laboratory experiences, because we do not consider the research evidence sufficient to support detailed policy prescriptions. A serious research agenda is required to build knowledge of how various types of laboratory experiences (within the context of science education) may contribute to specific science learning outcomes. Research partnerships may be the best mechanism to carry out this agenda, building the knowledge base for improvements in laboratory teaching and learning. Specifically, we suggest that teachers, researchers, scientists, and curriculum developers work together to answer the following questions. Addressing these questions will help to guide schools, education policy makers, and researchers in developing appropriate responses to the findings and conclusions in this report:

1. *Assessment of student learning in laboratory experiences*—What are the specific learning outcomes of laboratory experiences and what are the best methods for measuring these outcomes, both in the classroom and in large-scale assessments?

2. *Effective teaching and learning in laboratory experiences*—What forms of laboratory experiences are most effective for advancing the desired learning outcomes of laboratory experiences? What kinds of curriculum can support teachers in students in progress toward these learning outcomes?

3. *Diverse populations of learners*—What are the teaching and learning processes by which laboratory experiences contribute to particular learning outcomes for diverse learners and different populations of students?

4. *School organization for effective laboratory teaching*—What organizational arrangements (state and district policy, funding priorities and resource allocation, professional development, textbooks, emerging technologies, and school and district leadership) support high-quality laboratory experiences most efficiently and effectively? What are the most effective ways to bring those organizational arrangements to scale?

5. *Continuing learning about laboratory experiences*—How can teachers and administrators learn to design and implement effective instructional sequences that integrate laboratory experiences for diverse students? What types of professional development are most effective to help administrators and teachers achieve this goal? How should laboratory professional development be sequenced within a teacher's career (from pre-service to expert teacher)?

The available research literature suggests that laboratory experiences will be more likely to help students attain science learning goals if they are designed with clear learning outcomes in mind, thoughtfully sequenced into the flow of classroom science instruction, and follow the other instructional design principles identified by the committee. These design principles can serve as a guide to research, development, selection, and implementation of high school science curricula. They can also guide improvements in the undergraduate science education of future teachers and professional development of current science teachers.

The committee envisions a future in which the role and value of high school science laboratory experiences are more completely understood. The state of the research knowledge base on laboratory experience is dismal but, even so, suggests that the laboratory experiences of most high school students are equally dismal. Improvements in current laboratory experiences can be made today using emerging knowledge. Documented disparities to access should be eliminated now.

Systematic accumulation of rigorous, relevant research results and best practices from the field will clarify the specific contributions of laboratory experiences to science education. Such a knowledge base must be integrated with an infrastructure that supports the dissemination and use of this knowledge to achieve coherent policy and practice.

Improving the quality of laboratory experiences available to U.S. high school students will require focused and sustained attention. By applying principles of instructional design derived from ongoing research, science educators can begin to more effectively integrate laboratory experiences into the science curriculum. The definition, goals, design principles, and findings of this report offer an organizing framework to begin the difficult work of designing laboratory experiences for the 21st century.

1

Introduction, History, and Definition of Laboratories

Key Points

- *Since laboratories were introduced in the late 1800s, the goals of high school science education have changed. Today, high school science education aims to provide scientific literacy for all as part of a liberal education and to prepare students for further study, work, and citizenship.*
- *Educators and researchers do not agree on the definition and goals of high school science laboratories or on their role in the high school science curriculum.*
- *The committee defines high school science laboratories as follows: laboratory experiences provide opportunities for students to interact directly with the material world (or with data drawn from the material world), using the tools, data collection techniques, models, and theories of science.*

Science laboratories have been part of high school education for two centuries, yet a clear articulation of their role in student learning of science remains elusive. This report evaluates the evidence about the role of laboratories in helping students attain science learning goals and discusses factors that currently limit science learning in high school laboratories. In this chapter, the committee presents its charge, reviews the history of science laboratories in U.S. high schools, defines laboratories, and outlines the organization of the report.

CHARGE TO THE COMMITTEE

In the National Science Foundation (NSF) Authorization Act of 2002 (P.L. 107-368, authorizing funding for fiscal years 2003-2007), Congress called on NSF to launch a secondary school systemic initiative. The initiative was to "promote scientific and technological literacy" and to "meet the mathematics and science needs of students at risk of not achieving State student academic achievement standards." Congress directed NSF to provide grants for such activities as "laboratory improvement and provision of instrumentation as part of a comprehensive program to enhance the quality of mathematics, science, engineering, and technology instruction" (P.L. 107-368, Section 8-E). In response, NSF turned to the National Research Council (NRC) of the National Academies. NSF requested that the NRC

> nominate a committee to review the status of and future directions for the role of high school science laboratories in promoting the teaching and learning of science for all students. This committee will guide the conduct of a study and author a consensus report that will provide guidance on the question of the role and purpose of high school science laboratories with an emphasis on future directions. . . . Among the questions that may guide these activities are:
>
> 1. What is the current state of science laboratories and what do we know about how they are used in high schools?
> 2. What examples or alternatives are there to traditional approaches to labs and what is the evidence base as to their effectiveness?
> 3. If labs in high school never existed (i.e., if they were to be planned and designed de novo), what would that experience look like now, given modern advances in the natural and learning sciences?
> 4. In what ways can the integration of technologies into the curriculum augment and extend a new vision of high school science labs? What is known about high school science labs based on principles of design?
> 5. How do the structures and policies of high schools (course scheduling, curricular design, textbook adoption, and resource deployment) influence the organization of science labs? What kinds of changes might be needed in the infrastructure of high schools to enhance the effectiveness of science labs?
> 6. What are the costs (e.g., financial, personnel, space, scheduling) associated with different models of high school science labs? How might a new vision of laboratory experiences for high school students influence those costs?

7. In what way does the growing interdisciplinary nature of the work of scientists help to shape discussions of laboratories as contexts in high school for science learning?

8. How do high school lab experiences align with both middle school and postsecondary education? How is the role of teaching labs changing in the nation's colleges and universities? Would a redesign of high school science labs enhance or limit articulation between high school and college-level science education?

The NRC convened the Committee on High School Science Laboratories: Role and Vision to address this charge.

SCOPE OF THE STUDY

The committee carried out its charge through an iterative process of gathering information, deliberating on it, identifying gaps and questions, gathering further information to fill these gaps, and holding further discussions. In the search for relevant information, the committee held three public fact-finding meetings, reviewed published reports and unpublished research, searched the Internet, and commissioned experts to prepare and present papers. At a fourth, private meeting, the committee intensely analyzed and discussed its findings and conclusions over the course of three days. Although the committee considered information from a variety of sources, its final report gives most weight to research published in peer-reviewed journals and books.

At an early stage in its deliberations, the committee chose to focus primarily on "the role of high school laboratories in promoting the teaching and learning of science for all students." The committee soon became frustrated by the limited research evidence on the role of laboratories in learning. To address one of many problems in the research evidence—a lack of agreement about what constitutes a laboratory and about the purposes of laboratory education—the committee commissioned a paper to analyze the alternative definitions and goals of laboratories.

The committee developed a concept map outlining the main themes of the study (see Figure 1-1) and organized the three fact-finding meetings to gather information on each of these themes. For example, reflecting the committee's focus on student learning ("how students learn science" on the concept map), all three fact-finding meetings included researchers who had developed innovative approaches to high school science laboratories. We also commissioned two experts to present papers reviewing available research on the role of laboratories in students' learning of science.

At the fact-finding meetings, some researchers presented evidence of student learning following exposure to sequences of instruction that included laboratory experiences; others provided data on how various technologies

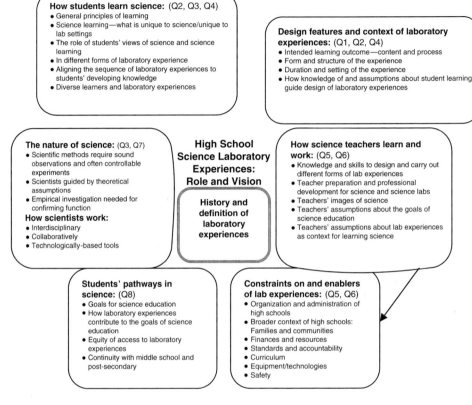

FIGURE 1-1 High school science laboratory experiences: Role and vision. Concept map with references to guiding questions in committee charge.

contribute to student learning in the laboratory. Responding to the congressional mandate to meet the mathematics and science needs of students at risk of not achieving state student academic achievement standards, the third fact-finding meeting included researchers who have studied laboratory teaching and learning among diverse students. Taken together, all of these activities enabled the committee to address questions 2, 3, and 4 of the charge.

The committee took several steps to ensure that the study reflected the current realities of science laboratories in U.S high schools, addressing the themes of "how science teachers learn and work" and "constraints and enablers of laboratory experiences" on the concept map. At the first fact-finding meeting, representatives of associations of scientists and science teachers described their efforts to help science teachers learn to lead effective laboratory activities. They noted constraints on laboratory learning, including poorly designed, overcrowded laboratory classrooms and inadequate preparation of science teachers. This first meeting also included a presentation about laboratory scheduling, supplies, and equipment drawn from a national survey of science teachers conducted in 2000. At the second fact-finding meeting, an architect spoke about the design of laboratory facilities, and a sociologist described how the organization of work and authority in schools may enable or constrain innovative approaches to laboratory teaching. Two meetings included panel discussions about laboratory teaching among groups of science teachers and school administrators. Through these presentations, review of additional literature, and internal discussions, the committee was able to respond to questions 1, 5, and 6 of the charge. The agendas for each fact-finding meeting, including the guiding questions that were sent to each presenter, appear in Appendix A.

The committee recognized that the question in its charge about the increasingly interdisciplinary nature of science (question 7) is important to the future of science and to high school science laboratories. In presentations and commissioned papers, several experts offered suggestions for how laboratory activities could be designed to more accurately reflect the work of scientists and to improve students' understanding of the way scientists work today. Based on our analysis of this information, the committee partially addresses this question from the perspective of how scientists conduct their work (in this chapter). The committee also identifies design principles for laboratory activities that may increase students' understanding of the nature of science (in Chapter 3). However, in order to maintain our focus on the key question of student learning in laboratories, the committee did not fully address question 7.

Another important question in the committee's charge (question 8) addresses the alignment of laboratory learning in middle school, high school, and undergraduate science education. Within the short time frame of this study, the committee focused on identifying, assembling, and analyzing the limited research available on high school science laboratories and did not attempt to do the same analysis for middle school and undergraduate science laboratories. However, this report does discuss several studies of student laboratory learning in middle school (see Chapter 3) and describes undergraduate science laboratories briefly in its analysis of the preparation of high school science teachers (see in Chapter 5). The committee thinks questions about the alignment of laboratory learning merit more sustained attention than was possible in this study.

During the course of our deliberations, other important questions emerged. For example, it is apparent that the scientific community is engaged in an array of efforts to strengthen teaching and learning in high school science laboratories, but little information is available on the extent of these efforts and on their effectiveness at enhancing student learning. As a result, we address the role of the scientific community in high school laboratories only briefly in Chapters 1 and 5. Another issue that arose over the course of this study is laboratory safety. We became convinced that laboratory safety is critical, but we did not fully analyze safety issues, which lay outside our charge. Finally, although engaging students in design or engineering laboratory activities appears to hold promising connections with science laboratory activities, the committee did not explore this possibility. Although all of these issues and questions are important, taking time and energy to address them would have deterred us from a central focus on the role of high school laboratories in promoting the teaching and learning of science for all students.

One important step in defining the scope of the study was to review the history of laboratories. Examining the history of laboratory education helped to illuminate persistent tensions, provided insight into approaches to be avoided in the future, and allowed the committee to more clearly frame key questions for the future.

HISTORY OF LABORATORY EDUCATION

The history of laboratories in U.S. high schools has been affected by changing views of the nature of science and by society's changing goals for science education. Between 1850 and the present, educators, scientists, and the public have, at different times, placed more or less emphasis on three sometimes-competing goals for school science education: (1) a theoretical emphasis, stressing the structure of scientific disciplines, the benefits of basic scientific research, and the importance of preparing young people for higher education in science; (2) an applied or practical emphasis, stressing high school students' ability to understand and apply the science and workings of everyday things; and (3) a liberal or contextual emphasis, stressing the historical development and cultural implications of science (Matthews, 1994). These changing goals have affected the nature and extent of laboratory education.

1850-1950

By the mid-19th century, British writers and philosophers had articulated a view of science as an inductive process (Mill, 1843; Whewell, 1840, 1858). They believed that scientists engaged in painstaking observation of nature to identify and accumulate facts, and only very cautiously did they draw conclusions from these facts to propose new theories. British and American scientists portrayed the newest scientific discoveries—such as the laws of thermodynamics and Darwin's theory of evolution—to an increasingly interested public as certain knowledge derived through well-established inductive methods. However, scientists and teachers made few efforts to teach students about these methods. High school and undergraduate science courses, like those in history and other subjects, were taught through lectures and textbooks, followed by rote memorization and recitation (Rudolph, 2005). Lecturers emphasized student knowledge of the facts, and science laboratories were not yet accepted as part of higher education. For example, when Benjamin Silliman set up the first chemistry laboratory at Yale in 1847, he paid rent to the college for use of the building and equipped it at his own expense (Whitman, 1898, p. 201). Few students were allowed into these laboratories, which were reserved for scientists' research, although some apparatus from the laboratory was occasionally brought into the lecture room for demonstrations.

During the 1880s, the situation changed rapidly. Influenced by the example of chemist Justus von Liebig in Germany, leading American universities embraced the German model. In this model, laboratories played a central role as the setting for faculty research and for advanced scientific study by students. Johns Hopkins University established itself as a research institution with student laboratories. Other leading colleges and universities followed suit, and high schools—which were just being established as educational institutions—soon began to create student science laboratories as well.

The primary goal of these early high school laboratories was to prepare students for higher science education in college and university laboratories. The National Education Association produced an influential report noting the "absolute necessity of laboratory work" in the high school science curriculum (National Education Association, 1894) in order to prepare students for undergraduate science studies. As demand for secondary school teachers trained in laboratory methods grew, colleges and universities began offering summer laboratory courses for teachers. In 1895, a zoology professor at Brown University described "large and increasing attendance at our summer schools," which focused on the dissection of cats and other animals (Bump, 1895, p. 260).

In these early years, American educators emphasized the theoretical, disciplinary goals of science education in order to prepare graduates for further science education. Because of this emphasis, high schools quickly embraced a detailed list of 40 physics experiments published by Harvard instructor Edwin Hall (Harvard University, 1889). The list outlined the experiments, procedures, and equipment necessary to successfully complete all 40 experiments as a condition of admission to study physics at Harvard. Scientific supply companies began selling complete sets of the required equipment to schools and successful completion of the exercises was soon required for admission to study physics at other colleges and universities (Rudolph, 2005).

At that time, most educators and scientists believed that participating in laboratory experiments would help students learn methods of accurate observation and inductive reasoning. However, the focus on prescribing specific experiments and procedures, illustrated by the embrace of the Harvard list, limited the effectiveness of early laboratory education. In the rush to specify laboratory experiments, procedures, and equipment, little attention had been paid to how students might learn from these experiences. Students were expected to simply absorb the methods of inductive reasoning by carrying out experiments according to prescribed procedures (Rudolph, 2005).

Between 1890 and 1910, as U.S. high schools expanded rapidly to absorb a huge influx of new students, a backlash began to develop against the prevailing approach to laboratory education. In a 1901 lecture at the New England Association of College and Secondary Schools, G. Stanley Hall, one of the first American psychologists, criticized high school physics education based on the Harvard list, saying that "boys of this age . . . want more dynamic physics" (Hall, 1901). Building on Hall's critique, University of Chicago physicist Charles Mann and other members of the Central Association for Science and Mathematics Teaching launched a complete overhaul of high school physics teaching. Mann and others attacked the "dry bones" of the Harvard experiments, calling for a high school physics curriculum with more personal and social relevance to students. One described lab work as "at best a very artificial means of supplying experiences upon which to build physical concepts" (Woodhull, 1909). Other educators argued that science teaching could be improved by providing more historical perspective, and high schools began reducing the number of laboratory exercises.

By 1910, a clear tension had emerged between those emphasizing laboratory experiments and reformers favoring an emphasis on interesting, practical science content in high school science. However, the focus on content also led to problems, as students became overwhelmed with "interesting" facts. New York's experience illustrates this tension. In 1890, the New York State Regents exam included questions asking students to design experiments (Champagne and Shiland, 2004). In 1905, the state introduced a new syllabus of physics topics. The content to be covered was so extensive that, over the course of a year, an average of half an hour could be devoted to each topic, virtually eliminating the possibility of including laboratory activities (Matthews, 1994). An outcry to return to more experimentation in science courses resulted, and in 1910 New York State instituted a requirement for 30 science laboratory sessions taking double periods in the syllabus for Regents science courses (courses preparing students for the New York State Regents examinations) (Champagne and Shiland, 2004).

In an influential speech to the American Association for the Advancement of Science (AAAS) in 1909, philosopher and educator John Dewey proposed a solution to the tension between advocates for more laboratory experimentation and advocates for science education emphasizing practical content. While criticizing science teaching focused strictly on covering large amounts of known content, Dewey also pointed to the flaws in rigid laboratory exercises: "A student may acquire laboratory methods as so much isolated and final stuff, just as he may so acquire material from a textbook. . . . Many a student had acquired dexterity and skill in laboratory methods without it ever occurring to him that they have anything to do with constructing beliefs that are alone worthy of the title of knowledge" (Dewey, 1910b). Dewey believed that people should leave school with some understanding of the kinds of evidence required to substantiate scientific beliefs. However, he never explicitly described his view of the process by which scientists develop and substantiate such evidence.

In 1910, Dewey wrote a short textbook aimed at helping teachers deal with students as individuals despite rapidly growing enrollments. He analyzed what he called "a complete act of thought," including five steps: (1) a felt difficulty, (2) its location and definition, (3) suggestion of possible solution, (4) development by reasoning of the bearing of the suggestion, and (5) further observation and experiment leading to its acceptance or rejection (Dewey, 1910a, pp. 68-78). Educators quickly misinterpreted these five steps as a description of the scientific method that could be applied to practical problems. In 1918, William Kilpatrick of Teachers College published a seminal article on the "project method," which used Dewey's five steps to address problems of everyday life. The article was eventually reprinted 60,000 times as reformers embraced the idea of engaging students with practical problems, while at the same time teaching them about what were seen as the methods of science (Rudolph, 2005).

During the 1920s, reform-minded teachers struggled to use the project method. Faced with ever-larger classes and state requirements for coverage of science content, they began to look for lists of specific projects that students could undertake, the procedures they could use, and the expected results. Soon, standardized lists of projects were published, and students who had previously been freed from rigid laboratory procedures were now engaged in rigid, specified projects, leading one writer to observe, "the project is little more than a new cloak for the inductive method" (Downing, 1919, p. 571).

Despite these unresolved tensions, laboratory education had become firmly established, and growing numbers of future high school teachers were instructed in teaching laboratory activities. For example, a 1925 textbook for preservice science teachers included a chapter titled "Place of Laboratory Work in the Teaching of Science" followed by three additional chapters on how to teach laboratory science (Brownell and Wade, 1925). Over the following decades, high school science education (including laboratory education) increasingly emphasized practical goals and the benefits of science in everyday life. During World War II, as scientists focused on federally funded research programs aimed at defense and public health needs, high school science education also emphasized applications of scientific knowledge (Rudolph, 2002).

1950-1975

Changing Goals of Science Education

Following World War II, the flood of "baby boomers" strained the physical and financial resources of public schools. Requests for increased taxes and bond issues led to increasing questions about public schooling. Some academics and policy makers began to criticize the "life adjustment" high school curriculum, which had been designed to meet adolescents' social, personal, and vocational needs. Instead, they called for a renewed emphasis on the academic disciplines. At the same time, the nation was shaken by the Soviet Union's explosion of an atomic bomb and the communist takeover of China. By the early 1950s, some federal policy makers began to view a more rigorous, academic high school science curriculum as critical to respond to the Soviet threat.

In 1956, physicist Jerrold Zacharias received a small grant from NSF to establish the Physical Science Study Committee (PSSC) in order to develop a curriculum focusing on physics as a scientific discipline. When the Union of Soviet Socialist Republics launched the space satellite Sputnik the following year, those who had argued that U.S. science education was not rigorous enough appeared vindicated, and a new era of science education began.

Although most historians believe that the overriding goal of the post-Sputnik science education reforms was to create a new generation of U.S. scientists and engineers capable of defending the nation from the Soviet Union, the actual goals were more complex and varied (Rudolph, 2002). Clearly, Congress, the president, and NSF were focused on the goal of preparing more scientists and engineers, as reflected in NSF director Alan Waterman's 1957 statement (National Science Foundation, 1957, pp. xv-xvi):

> Our schools and colleges are badly in need of modern science laboratories and laboratory, demonstration, and research equipment. Most important of all, we need more trained scientists and engineers in many special fields, and especially very many more competent, fully trained teachers of science, notably in our secondary schools. Undoubtedly, by a determined campaign, we can accomplish these ends in our traditional way, but how soon? The process is usually a lengthy one, and there is no time to be lost. Therefore, the pressing question is how quickly can our people act to accomplish these things?

The scientists, however, had another agenda. Over the course of World War II, their research had become increasingly dependent on federal funding and influenced by federal needs. In physics, for example, federally funded efforts to develop nuclear weapons led research to focus increasingly at the atomic level. In order to maintain public funding while reducing unwanted public pressure on research directions, the scientists sought to use curriculum redesign as a way to build the public's faith in the expertise of professional scientists (Rudolph, 2002). They wanted to emphasize the humanistic aspects of science, portraying science as an essential element in a broad liberal education. Some scientists sought to reach not only the select group who might become future scientists but also a slightly larger group of elite, mostly white male students who would be future leaders in government and business. They hoped to help these students appreciate the empirical grounding of scientific knowledge and to value and appreciate the role of science in society (Rudolph, 2002).

Changing Views of the Nature of Science

While this shift in the goals of science education was taking place, historians and philosophers were proposing new views of science. In 1958, British chemist Michael Polanyi questioned the ideal of scientific detachment and objectivity, arguing that scientific discovery relies on the personal participation and the creative, original thoughts of scientists (Polanyi, 1958). In the United States, geneticist and science educator Joseph Schwab suggested that scientific methods were specific to each discipline and that all scientific "inquiry" (his term for scientific research) was guided by the current theories and concepts within the discipline (Schwab, 1964). Publication of *The Structure of Scientific Revolutions* (Kuhn, 1962) a few years later fueled the debate about whether science was truly rational, and whether theory or observation was more important to the scientific enterprise. Over time, this debate subsided, as historians and philosophers of science came to focus on the process of scientific discovery. Increasingly, they recognized that this process involves deductive reasoning (developing inferences from known scientific principles and theories) as well as inductive reasoning (proceeding from particular observations to reach more general theories or conclusions).

Development of New Science Curricula

Although these changing views of the nature of science later led to changes in science education, they had little influence in the immediate aftermath of Sputnik. With NSF support, scientists led a flurry of curriculum development over the next three decades (Matthews, 1994). In addition to the physics text developed by the PSSC, the Biological Sciences Curriculum Study (BSCS) created biology curricula, the Chemical Education Materials group created chemistry materials, and groups of physicists created Introductory Physical Science and Project Physics. By 1975, NSF supported 28 science curriculum reform projects.

By 1977 over 60 percent of school districts had adopted at least one of the new curricula (Rudolph, 2002). The PSSC program employed high school teachers to train their peers in how to use the curriculum, reaching over half of all high school physics teachers by the late 1960s. However, due to implementation problems that we discuss further below, most schools soon shifted to other texts, and the federal goal of attracting a larger proportion of students to undergraduate science was not achieved (Linn, 1997).

Dissemination of the NSF-funded curriculum development efforts was limited by several weaknesses. Some curriculum developers tried to "teacher proof" their curricula, providing detailed texts, teacher guides, and filmstrips designed to ensure that students faithfully carried out the experiments as intended (Matthews, 1994). Physics teacher and curriculum developer Arnold Arons attributed the limited implementation of most of the NSF-funded curricula to lack of logistical support for science teachers and inadequate teacher training, since "curricular materials, however skilful and imaginative, cannot 'teach themselves'" (Arons, 1983, p. 117). Case studies showed that schools were slow to change in response to the new curricula and highlighted the central role of the teacher in carrying them out (Stake and Easley, 1978). In his analysis of Project Physics, Welch concluded that the new curriculum accounted for only 5 percent of the variance in student achievement, while other factors, such as teacher effectiveness, student ability, and time on task, played a larger role (Welch, 1979).

Despite their limited diffusion, the new curricula pioneered important new approaches to science education, including elevating the role of laboratory activities in order to help students understand the nature of modern scientific research (Rudolph, 2002). For example, in the PSSC curriculum, Massachusetts Institute of Technology physicist Jerrold Zacharias coordinated laboratory activities with the textbook in order to deepen students' understanding of the links between theory and experiments. As part of that curriculum, students experimented with a ripple tank, generating wave patterns in water in order to gain understanding of wave models of light. A new definition of the scientific laboratory informed these efforts. The PSSC text explained that a "laboratory" was a way of thinking about scientific investigations—an intellectual process rather than a building with specialized equipment (Rudolph, 2002, p. 131).

The new approach to using laboratory experiences was also apparent in the Science Curriculum Improvement Study. The study group drew on the developmental psychology of Jean Piaget to integrate laboratory experiences with other forms of instruction in a "learning cycle" (Atkin and Karplus, 1962). The learning cycle included (1) exploration of a concept, often through a laboratory experiment; (2) conceptual invention, in which the student or

TABLE 1-1 New Approaches Included in Post-Sputnik Science Curricula

	New Post-Sputnik Curricula	Traditional Science Curricula
Time of development	After 1955	Before 1955
Emphasis	Nature, structure, processes of science	Knowledge of scientific facts, laws, theories, applications
Role of laboratories	Integrated into the class routine	Secondary applications of concepts previously covered
Goals for students	Higher cognitive skills, appreciation of science	

SOURCE: Shymansky, Kyle, and Alport (1983). Reprinted with permission of Wiley-Liss, Inc., a subsidiary of John Wiley & Sons, Inc.

teacher (or both) derived the concept from the experimental data, usually during a classroom discussion; and (3) concept application in which the student applied the concept (Karplus and Their, 1967). Evaluations of the instructional materials, which were targeted to elementary school students, revealed that they were more successful than traditional forms of science instruction at enhancing students' understanding of science concepts, their understanding of the processes of science, and their positive attitudes toward science (Abraham, 1998). Subsequently, the learning cycle approach was applied to development of science curricula for high school and undergraduate students. Research into these more recent curricula confirms that "merely providing students with hands-on laboratory experiences is not by itself enough" (Abraham, 1998, p. 520) to motivate and help them understand science concepts and the nature of science.

In sum, the new approach of integrating laboratory experiences represented a marked change from earlier science education. In contrast to earlier curricula, which included laboratory experiences as secondary applications of concepts previously addressed by the teacher, the new curricula integrated laboratory activities into class routines in order to emphasize the nature and processes of science (Shymansky, Kyle, and Alport, 1983; see Table 1-1). Large meta-analyses of evaluations of the post-Sputnik curricula (Shymansky et al., 1983; Shymansky, Hedges, and Woodworth, 1990) found they were more effective than the traditional curriculum in boosting students' science achievement and interest in science. As we discuss in Chapter 3, current designs of science curricula that integrate laboratory experiences into ongoing classroom instruction have proven effective in enhancing students' science achievement and interest in science.

Discovery Learning and Inquiry

One offshoot of the curriculum development efforts in the 1960s and 1970s was the development of an approach to science learning termed "discovery learning." In 1959, Harvard cognitive psychologist Jerome Bruner began to develop his ideas about discovery learning as director of an NRC committee convened to evaluate the new NSF-funded curricula. In a book drawing in part on that experience, Bruner suggested that young students are active problem solvers, ready and motivated to learn science by their natural interest in the material world (Bruner, 1960). He argued that children should not be taught isolated science facts, but rather should be helped to discover the structures, or underlying concepts and theories, of science. Bruner's emphasis on helping students to understand the *theoretical* structures of the scientific disciplines became confounded with the idea of engaging students with the *physical* structures of natural phenomena in the laboratory (Matthews, 1994). Developers of NSF-funded curricula embraced this interpretation of Bruner's ideas, as it leant support to their emphasis on laboratory activities.

On the basis of his observation that scientific knowledge was changing rapidly through large-scale research and development during this postwar period, Joseph Schwab advocated the closely related idea of an "inquiry approach" to science education (Rudolph, 2003). In a seminal article, Schwab argued against teaching science facts, which he termed a "rhetoric of conclusions" (Schwab, 1962, p. 25). Instead, he proposed that teachers engage students with materials that would motivate them to learn about natural phenomena through inquiry while also learning about some of the strengths and weaknesses of the processes of scientific inquiry. He developed a framework to describe the inquiry approach in a biology laboratory. At the highest level of inquiry, the student simply confronts the "raw phenomenon" (Schwab, 1962, p. 55) with no guidance. At the other end of the spectrum, biology students would experience low levels of inquiry, or none at all, if the laboratory manual provides the problem to be investigated, the methods to address the problem, and the solutions. When Herron applied Schwab's framework to analyze the laboratory manuals included in the PSSC and the BSCS curricula, he found that most of the manuals provided extensive guidance to students and thus did not follow the inquiry approach (Herron, 1971).

The NRC defines inquiry somewhat differently in the *National Science Education Standards*. Rather than using "inquiry" as an indicator of the amount of guidance provided to students, the NRC described inquiry as encompassing both "the diverse ways in which scientists study the natural world" (National Research Council, 1996, p. 23) and also students' activities that support the learning of science concepts and the processes of science. In the NRC definition, student inquiry may include reading about known scientific theories and ideas, posing questions, planning investigations, making observations, using tools to gather and analyze data, proposing explanations, reviewing known theories and concepts in light of empirical data, and communicating the results. The *Standards* caution that emphasizing inquiry does not mean relying on a single approach to science teaching, suggesting that teachers use a variety of strategies, including reading, laboratory activities, and other approaches to help students learn science (National Research Council, 1996).

Diversity in Schools

During the 1950s, as some scientists developed new science curricula for teaching a small group of mostly white male students, other Americans were much more concerned about the weak quality of racially segregated schools for black children. In 1954, the Supreme Court ruled unanimously that the Topeka, Kansas Board of Education was in violation of the U.S. Constitution because it provided black students with "separate but equal" education. Schools in both the North and the South changed dramatically as formerly all-white schools were integrated. Following the example of the civil rights movement, in the 1970s and the 1980s the women's liberation movement sought improved education and employment opportunities for girls and women, including opportunities in science. In response, some educators began to seek ways to improve science education for all students, regardless of their race or gender.

1975 to Present

By 1975, the United States had put a man on the moon, concerns about the "space race" had subsided, and substantial NSF funding for science education reform ended. These changes, together with increased concern for equity in science education, heralded a shift in society's goals for science education. Science educators became less focused on the goal of disciplinary knowledge for science specialists and began to place greater emphasis on a liberal, humanistic view of science education.

Many of the tensions evident in the first 100 years of U.S. high school laboratories have continued over the past 30 years. Scientists, educators, and policy makers continue to disagree about the nature of science, the goals of science education, and the role of the curriculum and the teacher in student learning. Within this larger dialogue, debate about the value of laboratory activities continues.

Changing Goals for Science Education

National reports issued during the 1980s and 1990s illustrate new views of the nature of science and increased emphasis on liberal goals for science education. In *Science for All Americans*, the AAAS advocated the achievement of scientific literacy by all U.S. high school students, in order to increase their awareness and understanding of science and the natural world and to develop their ability to think scientifically (American Association for the Advancement of Science, 1989). This seminal report described science as tentative (striving toward objectivity within the constraints of human fallibility) and as a social enterprise, while also discussing the durability of scientific theories, the importance of logical reasoning, and the lack of a single scientific method. In the ongoing debate about the coverage of science content, the AAAS took the position that "curricula must be changed to reduce the sheer amount of material covered" (American Association for the Advancement of Science, 1989, p. 5). Four years later, the AAAS published *Benchmarks for Science Literacy*, which identified expected competencies at each school grade level in each of the earlier report's 10 areas of scientific literacy (American Association for the Advancement of Science, 1993).

The NRC's *National Science Education Standards* (National Research Council, 1996) built on the AAAS reports, opening with the statement: "This nation has established as a goal that all students should achieve scientific literacy" (p. ix). The NRC proposed national science standards for high school students designed to help all students develop (1) abilities necessary to do scientific inquiry and (2) understandings about scientific inquiry (National Research Council, 1996, p. 173).

In the standards, the NRC suggested a new approach to laboratories that went beyond simply engaging students in experiments. The NRC explicitly recognized that laboratory investigations should be learning experiences, stating that high school students must "actively participate in scientific investigations, and . . . use the cognitive and manipulative skills associated with the formulation of scientific explanations" (National Research Council, 1996, p. 173).

According to the standards, regardless of the scientific investigation performed, students must use evidence, apply logic, and construct an argument for their proposed explanations. These standards emphasize the importance of creating scientific arguments and explanations for observations made in the laboratory.

While most educators, scientists, and policy makers now agree that scientific literacy for all students is the primary goal of high school science education, the secondary goals of preparing the future scientific and technical workforce and including science as an essential part of a broad liberal education remain important. In 2004, the NSF National Science Board released a report describing a "troubling decline" in the number of U.S. citizens training to become scientists and engineers at a time when many current scientists and engineers are soon to retire. NSF called for improvements in science education to reverse these trends, which "threaten the economic welfare and security of our country" (National Science Foundation, 2004, p. 1). Another recent study found that secure, well-paying jobs that do not require postsecondary education nonetheless require abilities that may be developed in science laboratories. These include the ability to use inductive and deductive reasoning to arrive at valid conclusions; distinguish among facts and opinions; identify false premises in an argument; and use mathematics to solve problems (Achieve, 2004).

Achieving the goal of scientific literacy for all students, as well as motivating some students to study further in science, may require diverse approaches for the increasingly diverse body of science students, as we discuss in Chapter 2.

Changing Role of Teachers and Curriculum

Over the past 20 years, science educators have increasingly recognized the complementary roles of curriculum and teachers in helping students learn science. Both evaluations of NSF-funded curricula from the 1960s and more recent research on science learning have highlighted the important role of the teacher in helping students learn through laboratory activities. Cognitive psychologists and science educators have found that the teacher's expectations, interventions, and actions can help students develop understanding of scientific concepts and ideas (Driver, 1995; Penner, Lehrer, and Schauble, 1998; Roth and Roychoudhury, 1993). In response to this growing awareness, some school districts and institutions of higher education have made efforts to improve laboratory education for current teachers as well as to improve the undergraduate education of future teachers (National Research Council, 2001).

In the early 1980s, NSF began again to fund the development of laboratory-centered high school science curricula. Today, several publishers offer comprehensive packages developed with NSF support, including textbooks, teacher guides, and laboratory materials (and, in some cases, videos and web sites). In 2001, one earth science curriculum, five physical science curricula, five life science curricula, and six integrated science curricula were available for sale, while several others in various science disciplines were still under development (Biological Sciences Curriculum Study, 2001). In contrast to the curriculum development approach of the 1960s, teachers have played an important role in developing and field-testing these newer curricula and in designing the teacher professional development courses that accompany most of them. However, as in the 1960s and 1970s, only a few of these NSF-funded curricula have been widely adopted. Private publishers have also developed a multitude of new textbooks, laboratory manuals, and laboratory equipment kits in response to the national education standards and the growing national concern about scientific literacy. Nevertheless, most schools today use science curricula that have not been developed, field-tested, or refined on the basis of specific education research (see Chapter 2).

CURRENT DEBATES

Clearly, the United States needs high school graduates with scientific literacy—both to meet the economy's need for skilled workers and future scientists and to develop the scientific habits of mind that can help citizens in their everyday lives. Science is also important as part of a liberal high school education that conveys an important aspect of modern culture. However, the value of laboratory experiences in meeting these national goals has not been clearly established.

Researchers agree neither on the desired learning outcomes of laboratory experiences nor on whether those outcomes are attained. For example, on the basis of a 1978 review of over 80 studies, Bates concluded that there was no conclusive answer to the question, "What does the laboratory accomplish that could not be accomplished as well by less expensive and less time-consuming alternatives?" (Bates, 1978, p. 75). Some experts have suggested that the only contribution of laboratories lies in helping students develop skills in manipulating equipment and acquiring a feel for phenomena but that laboratories cannot help students understand science concepts (Woolnough, 1983; Klopfer, 1990). Others, however, argue that laboratory experiences have the potential to help students understand complex science concepts, but the potential has not been realized (Tobin, 1990; Gunstone and Champagne, 1990).

These debates in the research are reflected in practice. On one hand, most states and school districts continue to invest in laboratory facilities and equipment, many undergraduate institutions require completion of laboratory courses to qualify for admission, and some states require completion of science laboratory courses as a condition of high school graduation. On the other hand, in early 2004, the California Department of Education considered draft criteria for the evaluation of science instructional materials that reflected skepticism about the value of laboratory experiences or other hands-on learning activities. The proposed criteria would have required materials to demonstrate that the state science standards could be comprehensively covered with hands-on activities composing no more than 20 to 25 percent of instructional time (Linn, 2004). However, in response to opposition, the criteria were changed to require that the instructional materials would comprehensively cover the California science standards with "hands-on activities composing *at least* 20 to 25 percent of the science instructional program" (California Department of Education, 2004, p. 4, italics added).

The growing variety in laboratory experiences—which may be designed to achieve a variety of different learning outcomes—poses a challenge to resolving these debates. In a recent review of the literature, Hofstein and Lunetta (2004, p. 46) call attention to this variety:

> The assumption that laboratory experiences help students understand materials, phenomena, concepts, models and relationships, almost independent of the nature of the laboratory experience, continues to be widespread in spite of sparse data from carefully designed and conducted studies.

As a first step toward understanding the nature of the laboratory experience, the committee developed a definition and a typology of high school science laboratory experiences.

DEFINITION OF LABORATORY EXPERIENCES

Rapid developments in science, technology, and cognitive research have made the traditional definition of science laboratories—as rooms in which students use special equipment to carry out well-defined procedures—obsolete. The committee gathered information on a wide variety of approaches to laboratory education, arriving at the term "laboratory experiences" to describe teaching and learning that may take place in a laboratory room or in other settings:

> Laboratory experiences provide opportunities for students to interact directly with the material world (or with data drawn from the material world), using the tools, data collection techniques, models, and theories of science.

This definition includes the following student activities:

- Physical manipulation of the real-world substances or systems under investigation. This may include such activities as chemistry experiments, plant or animal dissections in biology, and investigation of rocks or minerals for identification in earth science.
- Interaction with simulations. Physical models have been used throughout the history of science teaching (Lunetta, 1998). Today, students can work

with computerized models, or simulations, representing aspects of natural phenomena that cannot be observed directly, because they are very large, very small, very slow, very fast, or very complex. Using simulations, students may model the interaction of molecules in chemistry or manipulate models of cells, animal or plant systems, wave motion, weather patterns, or geological formations.

- Interaction with data drawn from the real world. Students may interact with real-world data that are obtained and represented in a variety of forms. For example, they may study photographs to examine characteristics of the moon or other heavenly bodies or analyze emission and absorption spectra in the light from stars. Data may be incorporated in films, DVDs, computer programs, or other formats.
- Access to large databases. In many fields of science, researchers have arranged for empirical data to be normalized and aggregated—for example, genome databases, astronomy image collections, databases of climatic events over long time periods, biological field observations. With the help of the Internet, some students sitting in science class can now access these authentic and timely scientific data. Students can manipulate and analyze these data drawn from the real world in new forms of laboratory experiences (Bell, 2005).
- Remote access to scientific instruments and observations. A few classrooms around the nation experience laboratory activities enabled by Internet links to remote instruments. Some students and teachers study insects by accessing and controlling an environmental scanning electron microscope (Thakkar et al., 2000), while others control automated telescopes (Gould, 2004).

Although we include all of these types of direct and indirect interaction with the material world in this definition, it does not include student manipulation or analysis of data created by a teacher to replace or substitute for direct interaction with the material world. For example, if a physics teacher presented students with a constructed data set on the weight and required pulling force for boxes pulled across desks with different surfaces, asking the students to analyze these data, the students' problem-solving activity would not constitute a laboratory experience according to the committee's definition.

Previous Definitions of Laboratories

In developing its definition, the committee reviewed previous definitions of student laboratories. Hegarty-Hazel (1990, p. 4) defined laboratory work as:

> a form of practical work taking place in a purposely assigned environment where students engage in planned learning experiences . . . [and] interact

with materials to observe and understand phenomena (Some forms of practical work such as field trips are thus excluded).

Lunetta defined laboratories as "experiences in school settings in which students interact with materials to observe and understand the natural world" (Lunetta, 1998, p. 249). However, these definitions include only students' direct interactions with natural phenomena, whereas we include both such direct interactions and also student interactions with data drawn from the material world. In addition, these earlier definitions confine laboratory experiences to schools or other "purposely assigned environments," but our definition encompasses student observation and manipulation of natural phenomena in a variety of settings, including science museums and science centers, school gardens, local streams, or nearby geological formations. The committee's definition also includes students who work as interns in research laboratories, after school or during the summer months. All of these experiences, as well as those that take place in traditional school science laboratories, are included in our definition of laboratory experiences.

Variety in Laboratory Experiences

Both the preceding review of the history of laboratories and the committee's review of the evidence of student learning in laboratories reveal the limitations of engaging students in replicating the work of scientists. It has become increasingly clear that it is not realistic to expect students to arrive at accepted scientific concepts and ideas by simply experiencing some aspects of scientific research (Millar, 2004). While recognizing these limitations, the committee thinks that laboratory experiences should at least partially reflect the range of activities involved in real scientific research. Providing students with opportunities to participate in a range of scientific activities represents a step toward achieving the learning goals of laboratories identified in Chapter 3.[1]

Historians and philosophers of science now recognize that the well-ordered scientific method taught in many high school classes does not exist. Scientists' empirical research in the laboratory or the field is one part of a larger process that may include reading and attending conferences to stay abreast of current developments in the discipline and to present work in progress. As Schwab recognized (1964), the "structure" of current theories and concepts in a discipline acts as a guide to further empirical research. The work of scientists may include formulating research questions, generat-

---

[1]The goals of laboratory learning are unlikely to be reached, regardless of what type of laboratory experience is provided, unless the experience is well integrated into a coherent stream of science instruction, incorporates other design elements, and is led by a knowledgeable teacher, as discussed in Chapters 3 and 4.

ing alternative hypotheses, designing and conducting investigations, and building and revising models to explain the results of their investigations. The process of evaluating and revising models may generate new questions and new investigations (see Table 1-2). Recent studies of science indicate that scientists' interactions with their peers, particularly their response to questions from other scientists, as well as their use of analogies in formulating hypotheses and solving problems, and their responses to unexplained results, all influence their success in making discoveries (Dunbar, 2000). Some scientists concentrate their efforts on developing theory, reading, or conducting thought experiments, while others specialize in direct interactions with the material world (Bell, 2005).

Student laboratory experiences that reflect these aspects of the work of scientists would include learning about the most current concepts and theories through reading, lectures, or discussions; formulating questions; designing and carrying out investigations; creating and revising explanatory models; and presenting their evolving ideas and scientific arguments to others for discussion and evaluation (see Table 1-3).

Currently, however, most high schools provide a narrow range of laboratory activities, engaging students primarily in using tools to make observations and gather data, often in order to verify established scientific knowledge. Students rarely have opportunities to formulate research questions or to build and revise explanatory models (see Chapter 4).

ORGANIZATION OF THE REPORT

The ability of high school science laboratories to help improve all citizens' understanding and appreciation of science and prepare the next generation of scientists and engineers is affected by the context in which laboratory experiences take place. Laboratory experiences do not take place in isolation, but are part of the larger fabric of students' experiences during their high school years. Following this introduction, Chapter 2 describes recent trends in U.S. science education and policies influencing science education, including laboratory experiences. In Chapter 3 we turn to a review of available evidence on student learning in laboratories and identify principles for design of effective laboratory learning environments. Chapter 4 describes current laboratory experiences in U.S. high schools, and Chapter 5 discusses teacher and school readiness for laboratory experiences. In Chapter 6, we describe the current state of laboratory facilities, equipment, and safety. Finally, in Chapter 7, we present our conclusions and an agenda designed to help laboratory experiences fulfill their potential role in the high school science curriculum.

TABLE 1-2 A Typology of Scientists' Activities

Type of Activity	Explanation
Posing a research question	One of the most difficult steps in science is to define a research question. A researchable question may arise out of analysis of data collected, or be based on already known scientific theories and laws, or both. While the initial question is important as a goal to guide the study, flexibility is also valuable. Scientists who respond to unexpected results (that do not fit current theories about the phenomena) by conducting further research to try to explain them are more likely to make discoveries than scientists whose goal is to find evidence consistent with their current knowledge (Dunbar, 1993, 2000; Merton and Barber, 2004).
Formulating hypotheses	Scientists sometimes generate one or more competing hypotheses related to a research question. However, not all scientific research is hypothesis-driven. The human genome project is an example of bulk data collection not driven by a hypothesis (Davies, 2001).
Designing investigations	Scientists design investigations—which may involve experimental or observational methods—to answer their research questions. Investigations may be designed to test one or more competing hypotheses.
Making observations, gathering, and analyzing data	Observing natural phenomena is often an essential part of a research project. Scientists use a variety of tools and procedures to make observations and gather data, searching for patterns and possible cause-and-effect relationships that may be studied further. Observations may be guided by theory, may be designed to test a hypothesis, or may explore unknown phenomena (Duschl, 2004).
Building or revising scientific models	Although modeling scientific phenomena has always been a central practice of science, it has only been recognized as a driving force in generating scientific knowledge over the past 50 years (Duschl, 2004). Scientists draw on their imagination and existing knowledge as they interpret data in order to develop explanatory models or theories (Driver et al., 1996). These models serve as tentative explanations for observations, subject to revision based on further observations or further study of known scientific principles or theories.
Evaluating, testing or verifying models	One of the defining characteristics of science is that the evidence, methods, and assumptions used to arrive at a proposed discovery are described and publicly disclosed so that other scientists can judge their validity (Hull, 1988; Longino, 1990, 1994). In one recent example, astronomers at the Green Bank radio telescope in West Virginia identified glycoaldehyde, a building block of DNA and RNA, in an extremely cold area of the Milky Way (Hollis et al., 2004). The discovery of this substance in an area of the galaxy where comets form suggests the possibility that the ingredients necessary to create life might have been carried to Earth by a comet billions of years ago. In a news report of the discovery, the director of the Arizona Radio Observatory, who had criticized the Green Bank astronomers for not being thorough enough, said her students had replicated the Green Bank observations (Gugliotta, 2004, p. A7).

TABLE 1-3 A Typology of School Laboratory Experiences

Type of Laboratory Experience	Description
Posing a research question	Formulating a testable question can be a great challenge for high school students. Some laboratory experiences may engage students in formulating and assessing the importance of alternative questions.
Using laboratory tools and procedures	Some laboratory experiences may be designed primarily to develop students' skills in making measurements and safely and correctly handling materials and equipment (Lunetta, 1998). These "prelab" exercises can help reduce errors and increase safety in subsequent laboratory experiences (Millar, 2004).
Formulating hypotheses	Like formulating a research question, formulating alternative hypotheses is challenging for high school students. However, some new curricula have led to improvement in formulating hypotheses (see Chapter 3).
Designing investigations	Laboratory experiences integrated with other forms of instruction and explicitly designed with this goal in mind can help students learn to design investigations (White and Frederiksen, 1998).
Making observations, gathering, and analyzing data	Science teachers may engage students in laboratory activities that involve observing phenomena and in gathering, recording, and analyzing data in search of possible patterns or explanations.
Building or revising models	Laboratory experiences may engage students in interpreting data that they gather directly from the material world or data drawn from large scientific data sets in order to create, test, and refine models. Scientific modeling is a core element in several innovative laboratory-centered science curricula that appear to enhance student learning (Bell, 2005).
Evaluating, testing, or verifying explanatory models (including known scientific theories and models)	Laboratory experiences may be designed to engage students in verifying scientific ideas that they have learned about through reading, lectures, or work with computer simulations. Such experiences can help students to understand accepted scientific concepts through their own direct experiences (Millar, 2004). However, verification laboratory activities are quite different from the activities of scientists who rigorously test a proposed scientific theory or discovery in order to defend, refute, or revise it.

SUMMARY

Since the late 19th century, high school students in the United States have carried out laboratory investigations as part of their science classes. Since that time, changes in science, education, and American society have influenced the role of laboratory experiences in the high school science curriculum. At the turn of the 20th century, high school science laboratory experiences were designed primarily to prepare a select group of young people for further scientific study at research universities. During the period between World War I and World War II, many high schools emphasized the more practical aspects of science, engaging students in laboratory projects related to daily life. In the 1950s and 1960s, science curricula were redesigned to integrate laboratory experiences into classroom instruction, with the goal of increasing public appreciation of science.

Policy makers, scientists, and educators agree that high school graduates today, more than ever, need a basic understanding of science and technology to function effectively in an increasingly complex, technological society. They seek to help students understand the nature of science and to develop both the inductive and deductive reasoning skills that scientists apply in their work. However, researchers and educators do not agree on how to define high school science laboratories or on their purposes, hampering the accumulation of evidence that might guide improvements in laboratory education. Gaps in the research and in capturing the knowledge of expert science teachers make it difficult to reach precise conclusions on the best approaches to laboratory teaching and learning.

In order to provide a focus for the study, the committee defines laboratory experiences as follows: laboratory experiences provide opportunities for students to interact directly with the material world (or with data drawn from the material world), using the tools, data collection techniques, models, and theories of science. This definition includes a variety of types of laboratory experiences, reflecting the range of activities that scientists engage in. The following chapters discuss the educational context; laboratory experiences and student learning; current laboratory experiences, teacher and school readiness, facilities, equipment, and safety; and laboratory experiences for the 21st century.

REFERENCES

Abraham, M.R. (1998). The learning cycle approach as a strategy for instruction in science. In B.J. Fraser and K.G. Tobin (Eds.), *International handbook of science education.* London, England: Kluwer Academic.

Achieve. (2004). *Ready or not: Creating a high school diploma that counts.* (The American Diploma Project.) Washington, DC: Author.

American Association for the Advancement of Science. (1989). *Science for all Americans*. Washington, DC: Author.

American Association for the Advancement of Science. (1993). *Benchmarks for science literacy*. Washington, DC: Author.

Arons, A. (1983). Achieving wider scientific literacy. *Daedalus, 112*(2), 91-122.

Atkin, J.M., and Karplus, R. (1962). Discovery or invention? *The Science Teacher, 29*, 45-51.

Bates, G.R. (1978). The role of the laboratory in science teaching. *School Science and Mathematics, 83*, 165-169.

Bell, P. (2005). *The school science laboratory: Considerations of learning, technology, and scientific practice*. Paper prepared for the Committee on High School Science Laboratories: Role and Vision. Available at: http://www7.nationalacademies.org/bose/July_12-13_2004_High_School_Labs_Meeting_Agenda.html.

Biological Sciences Curriculum Study. (2001). *Profiles in science: A guide to NSF-funded high school instructional materials*. Colorado Springs, CO: Author. Available at: http://www.bscs.org/page.asp?id=Professional_ Development l Resources l Profiles_In_Science [accessed Sept. 2004].

Brownell, H., and Wade, F.B. (1925). *The teaching of science and the science teacher*. New York: Century Company.

Bruner, J.S. (1960). *The process of education*. New York: Vintage.

Bump, H.C. (1895). Laboratory teaching of large classes—Zoology. [Letter to the editor]. *Science, New Series, 1*(10), 260.

California Department of Education. (2004). *Criteria for evaluating instructional materials in science, kindergarten through grade eight*. Available at: http://www.cde.ca.gov/ci/sc/cf/documents/scicriteria04.pdf [accessed Nov. 2004].

Champagne, A., and Shiland, T. (2004). *Large-scale assessment and the high school science laboratory*. Presentation to the Committee on High School Science Laboratories: Role and Vision, June 3-4, National Research Council, Washington, DC. Available at: http://www7.nationalacademies.org/bose/June_3-4_2004_High_School_Labs_Meeting_Agenda.html [accessed Jan. 2004].

Davies, K. (2001). *Cracking the genome: Inside the race to unlock human DNA*. New York: Free Press.

Dewey, J. (1910a). *How we think*. Boston, MA: D.C. Heath.

Dewey, J. (1910b). Science as subject-matter and method. *Science, New Series, 31*, 122, 125.

Downing, E.R. (1919). The scientific method and the problems of school science. *School and Society, 10*, 571.

Driver, R. (1995). Constructivist approaches to science teaching. In L.P. Steffe and J. Gale (Eds.), *Constructivism in education* (pp. 385-400). Hillsdale, NJ: Lawrence Erlbaum.

Driver, R., Leach, J., Millar, R., and Scott, P. (1996). *Young people's images of science*. Buckingham, U.K.: Open University Press.

Dunbar, K. (1993). Concept discovery in a scientific domain. *Cognitive Science, 17*, 397-434.

Dunbar, K. (2000). How scientists think in the real world: Implications for science education. *Journal of Applied Developmental Psychology 21*(1), 49-58.

Duschl, R. (2004). *The HS lab experience: Reconsidering the role of evidence, explanation, and the language of science.* Paper prepared for the Committee on High School Science Laboratories: Role and Vision, July 12-13, National Research Council, Washington, DC. Available at: http://www7.nationalacademies.org/bose/July_12-13_2004_High_School_Labs_Meeting_Agenda.html [accessed Nov. 2004].

Gould, R. (2004). *About micro observatory.* Cambridge, MA: Harvard University. Available at: http://mo-www.harvard.edu/MicroObservatory/ [accessed Sept. 2004].

Gugliotta, G. (2004). Space sugar a clue to life's origins. *The Washington Post*, September 27.

Gunstone, R.F., and Champagne, A.B. (1990). Promoting conceptual change in the laboratory. In E. Hegarty-Hazel (Ed.), *The student laboratory and the science curriculum* (pp. 159-182). London, England: Routledge.

Hall, G.S. (1901). How far is the present high-school and early college training adapted to the nature and needs of adolescents? *School Review, 9,* 652.

Harvard University. (1889). *Descriptive list of elementary physical experiments intended for use in preparing students for Harvard College.* Cambridge, MA: Author.

Hegarty-Hazel, E. (1990). Overview. In E. Hegarty-Hazel (Ed.), *The student laboratory and the science curriculum.* London, England: Routledge.

Herron, M.D. (1971). The nature of scientific enquiry. *School Review, 79,* 171-212.

Hilosky, A., Sutman, F., and Schmuckler, J. (1998). Is laboratory-based instruction in beginning college-level chemistry worth the effort and expense? *Journal of Chemical Education, 75*(1), 103.

Hofstein, A., and Lunetta, V.N. (2004). The laboratory in science education: Foundations for the twenty-first century. *Science Education, 88,* 28-54.

Hollis, J.M., Jewell, P.R., Lovas, F.J., and Remijan, J. (2004, September 20). Green Bank telescope observations of interstellar glycolaldehyde: Low-temperature sugar. *Astrophysical Journal, 613,* L45-L48.

Hull, D. (1988). *Science as a process.* Chicago: University of Chicago Press.

Karplus, R., and Their, H.D. (1967). *A new look at elementary school science.* Chicago: Rand McNally.

Klopfer, L.E. (1990). Learning scientific enquiry in the student laboratory. In E. Hegarty-Hazel (Ed.), *The student laboratory and the science curriculum* (pp. 95-118). London, England: Routledge.

Kuhn, T.S. (1962). *The structure of scientific revolutions.* Chicago: University of Chicago Press.

Linn, M. (2004). *High school science laboratories: How can technology contribute?* Presentation to the Committee on High School Science Laboratories: Role and Vision. June 3-4, National Research Council, Washington, DC. Available at: http://www7.nationalacademies.org/bose/June_3-4_2004_High_School_Labs_Meeting_Agenda.html [accessed April 2005].

Linn, M.C. (1997). The role of the laboratory in science learning. *Elementary School Journal, 97*(4), 401-417.

Longino, H. (1990). *Science as social knowledge.* Princeton, NJ: Princeton University Press.

Longino, H. (1994). The fate of knowledge in social theories of science. In F.F. Schmitt (Ed.), *Socializing epistemology: The social dimensions of knowledge* (pp. 135-158). Lanham, MD: Rowman and Littlefield.

Lunetta, V. (1998). *The school science laboratory: Historical perspectives and contexts for contemporary teaching.* In B.J. Fraser and K.G. Tobin (Eds.), *International handbook of science education* (pp. 249-262). London, England: Kluwer Academic.

Matthews, M.R. (1994). *Science teaching: The role of history and philosophy of science.* New York: Routledge.

Merton, R.K., and Barber, E. (2004). *The travels and adventures of serendipity.* Princeton, NJ: Princeton University Press.

Mill, J.S. (1963). System of logic, ratiocinative and inductive. In J.M. Robson (Ed.), *Collected works of John Stuart Mill.* Toronto: University of Toronto Press (Original work published 1843).

Millar, R. (2004). *The role of practical work in the teaching and learning of science.* Paper prepared for the Committee on High School Science Laboratories: Role and Vision, June 3-4, National Research Council, Washington, DC. Available at: http://www7.nationalacademies.org/bose/June3-4_2004_High_School_Labs_Meeting_Agenda.html [accessed April 2005].

National Education Association. (1894). *Report of the Committee of Ten on secondary school studies.* New York: American Book Company.

National Research Council. (1996). *National science education standards.* National Committee on Science Education Standards and Assessment, Center for Science, Mathematics, and Engineering Education. Washington, DC: National Academy Press.

National Research Council. (2001). *Educating teachers of science, mathematics, and technology: New practices for the new millennium.* Committee on Science and Mathematics Teacher Preparation, Center for Education. Washington, DC: National Academy Press.

National Science Foundation. (1957). *Annual report.* Available at: http://www.nsf.gov/pubs/1957/annualreports/start.htm [accessed Nov. 2004].

National Science Foundation. (2004). *An emerging and critical problem of the science and engineering workforce.* Washington, DC: Author. Available at: http://www.nsf.gov/sbe/srs/nsb0407/start.htm [accessed Sept. 2003].

Penner, D.E., Lehrer, R., and Schauble, L. (1998). From physical models to biomechanics: A design based modeling approach. *Journal of the Learning Sciences, 7,* 429-449.

Polanyi, M. (1958). *Personal knowledge: Towards a post-critical philosophy.* London, England: Routledge.

Roth, W.M., and Roychoudhury, A. (1993). The development of science process skills in authentic contexts. *Journal of Research in Science Teaching, 30,* 127-152.

Rudolph, J.L. (2002). *Scientists in the classroom: The cold war reconstruction of American science education.* New York: Palgrave.

Rudolph, J.L. (2003). Portraying epistemology: School science in historical context. *Science Education, 87,* 64-79.

Rudolph, J.L. (2005). Epistemology for the masses: The origins of the "scientific method" in American schools. *History of Education Quarterly, 45*(2), 341-376.

Schwab, J.J. (1962). The teaching of science as enquiry. In P.F. Brandwein (Ed.), *The teaching of science* (pp. 1-103). Cambridge, MA: Harvard University Press.

Schwab, J.J. (1964). The structure of the natural sciences. In G.W. Ford and L. Pugno (Eds.), *The structure of knowledge and the curriculum* (pp. 31-49). Chicago: Rand McNally.

Shymansky, J.A., Hedges, L.B., and Woodworth, G. (1990). A re-assessment of the effects of inquiry-based science curriculum of the sixties on student achievement. *Journal of Research in Science Teaching, 20*, 387-404.

Shymansky, J.A., Kyle, W.C., and Alport, J.M. (1983). The effects of new science curricula on student performance. *Journal of Research in Science Teaching, 20*, 387-404.

Stake, R.E., and Easley, J.A. (1978). *Case studies in science education.* Center for Instructional Research and Curriculum Evaluation and Committee on Culture and Cognition. Urbana: University of Illinois at Urbana-Champagne.

Thakkar, U., Carragher, B., Carroll, L., Conway, C., Grosser, B., Kisseberth, N., Potter, C.S., Robinson, S., Sinn-Hanlon, J., Stone, D., and Weber, D. (2000). *Formative evaluation of Bugscope: A sustainable world wide laboratory for K-12.* Paper prepared for the annual meeting of the American Educational Research Association, Special Interest Group on Advanced Technologies for Learning, April 24-28, New Orleans, LA. Available at: http://bugscope.beckman.uiuc.edu/publications/index.htm#papers [accessed May 2005].

Tobin, K. 1990. Research on science laboratory activities: In pursuit of better questions and answers to improve learning. *School Science and Mathematics, 90*(5), 403-418.

Welch, W.W. (1979). Twenty years of science education development: A look back. *Review of Research in Education, 7*, 282-306.

Whewell, W. (1840). *Philosophy of the inductive sciences, founded upon their history.* London, England: John W. Parker.

Whewell, W. (1858). *The history of the inductive sciences, from the earliest to the present time* (3rd edition). New York: Appleton.

White, B.Y., and Frederiksen, J.R. (1998). Inquiry, modeling, and metacognition: Making science accessible to all students. *Cognition and Instruction, 16*(1), 3-118.

Whitman, F.P. (1898). The beginnings of laboratory teaching in America. *Science, New Series, 8*(190), 201-206.

Woodhull, J.F. (1909). How the public will solve the problems of science teaching. *School Science and Mathematics, 9*, 276.

Woolnough, B.E. (1983). Exercises, investigations and experiences. *Physics Education, 18*, 60-63.

2

The Education Context

Key Points

- *High school students' science achievement nationwide is not impressive and has not changed substantially in three decades.*
- *The national and state policy environment of science education is complex and interconnected. This complex landscape must be taken into account when reconsidering the role of laboratory experiences in high school science.*
- *Currently, policies influencing high school science education are not well aligned. Some policies and practices may constrain efforts to improve high school science laboratory experiences.*

This chapter provides an overview of current trends in science education and the key policies influencing science education. Understanding this context helps to reveal the dynamics that have shaped current high school laboratory experiences and may influence new approaches to high school science laboratories. The first section of this chapter describes current trends in science achievement and the changing student population. Against this backdrop, the second section identifies and briefly summarizes the array of national and state policies that shape science education. Whenever possible, in discussing each policy or program, we discuss its possible implications for laboratory experiences.

RECENT TRENDS IN U.S. SCIENCE EDUCATION

Policy makers, scientists, and educators have expressed growing concern about the nation's scientific literacy and the international competitiveness of its science and technology workforce. Here we describe recent trends in public understanding of science and in high school science education, which provides the foundational knowledge for the next generation of scientists and engineers.

Public Understanding of Science

Major science education reports published in the 1990s advocated broad scientific literacy for all students, including understanding of science concepts and of the processes and nature of science (American Association for the Advancement of Science, 1993; National Research Council, 1996). This type of broadly defined scientific literacy is an essential part of a liberal education. It can provide a strong knowledge base for high school graduates, preparing them for further science and technology education and also to work and live as citizens in an increasingly technological society. The available evidence suggests, however, that levels of scientific literacy are low and improving them is a slow and difficult process.

Northwestern University Professor Jon Miller has developed a systematic approach to defining and measuring public scientific literacy, in surveys conducted for the National Science Foundation (NSF) over the past two decades (Miller, 2004). Defining scientific literacy as the level of understanding required to read and comprehend the science section of *The New York Times, The Wall Street Journal,* or other comparable major newspapers and magazines, Miller uses several measures of this understanding (Miller, 2004).

The survey results reveal slight improvements in public understanding of science. The percentage of U.S. adults with a minimal understanding of the nature of scientific research (From your point of view, what does it mean to study something scientifically?) increased from 12 percent in 1957 to 21 percent in 1999. The fraction of U.S. adults who understood experimentation, including the reasons for using control and experimental groups in medical research, also grew, from 22 percent in 1993 to 35 percent in 1999.

Over the past 15 years, Miller and colleagues studied public understanding of four specific scientific concepts—molecules, DNA, radiation, and the nature of the universe—that often appear in news stories but are rarely explained in depth. They found that understanding of these concepts is slowly increasing but remains low. For example, the percentage of U.S.

adults who were able to provide a correct explanation of a molecule increased from 11 percent in 1997 to 13 percent in 1999. Compiling several of these measures into an overall measure of civic scientific literacy, Miller concluded that the percentage of U.S. adults who are scientifically literate grew from 10 percent in the late 1980s to 17 percent in 1999 (Miller, 2004). Despite this low level of scientific understanding, however, the surveys indicate that large majorities of adults continue to believe that scientific research is valuable for economic prosperity and quality of life.

Science Achievement in Secondary School

The low level of public understanding of science may be related to the quality of high school science education, including the laboratory experiences that are a part of that education. Results from three written tests—the National Assessment of Educational Progress (NAEP), the Trends in International Mathematics and Science Study (TIMSS), and the Organisation for Economic Co-Operation and Development's (OECD) Programme for International Student Assessment (PISA)—indicate little or no improvement in high school students' science achievement over the past 30 years.

Although high school science laboratories could potentially contribute to improvement in the science achievement of U.S. students, current large-scale achievement tests are not capable of measuring progress toward all of the goals of laboratory experiences. The committee identified several educational goals that high school laboratory experiences should help students attain. They include (1) enhancing mastery of science subject matter, (2) developing scientific reasoning, (3) understanding the complexity and ambiguity of empirical work, (4) developing practical skills, (5) understanding of the nature of science, (6) cultivating interest in science and in learning science, and (7) developing teamwork abilities (see Chapter 3 for a detailed discussion of each goal).

Results of National Science Achievement Tests

The NAEP includes two components—trend NAEP, which includes test items in science and other subjects, that has been administered many times over the past three decades, and NAEP subject-matter tests, which reflect current expectations for student learning in science and other subjects (National Research Council, 1999).

The performance of 17-year-olds on the science portion of the long-term trend NAEP provides some indication of the extent to which they have attained one of the goals of laboratory experiences—enhancing mastery of science subject matter. Although the test framework calls for measuring not only students' mastery of subject matter but also their ability to conduct

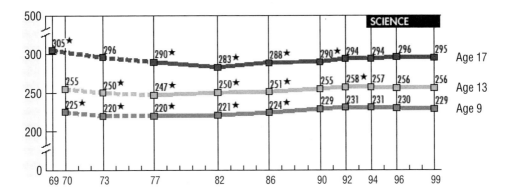

FIGURE 2-1 Long-term trends in average scale scores in science from NAEP.
NOTE: Dashed lines represent extrapolated data.
SOURCE: National Center for Education Statistics, National Assessment of Educational Progress (NAEP), 1999 Long-Term Trend Assessment.

inquiries and solve problems and their understanding of the nature of science (U.S. Department of Education, 2001), the test itself is composed entirely of selected-response items and emphasizes mastery of science subject matter. The long-term trend NAEP does not fully measure complex cognitive abilities that may be developed through laboratory experiences, such as the development of scientific reasoning and understanding of the complexity and ambiguity of empirical work (National Research Council, 1999).

The national average of scores of 17-year-olds on the science portion of the long-term trend NAEP assessment was lower in 1999 than 30 years earlier in 1969.[1] In contrast to this slight decline in 17-year-olds' scores, the average national scores of 13-year-olds and 9-year-olds increased very slightly over the 30-year period (see Figure 2-1). The overall trend for all ages suggests that U.S. students' science knowledge has not increased over the past three decades.

Student scores on the science portion of the long-term trend NAEP varied by racial/ethnic group and by gender. In 1999, white students had higher average scores than their black and Hispanic peers. Between 1970 and 1999, the gap between white and black students in science generally narrowed for 9- and 13-year-olds, but not for 17-year-olds, and the gap between white

---

[1] Students' scores on the long-term NAEP assessment are reported as average scale scores and also in terms of proficiency levels (basic, proficient, advanced). However, because a National Research Council committee that studied NAEP found the process for setting these proficiency levels to be flawed, they are not reported here (National Research Council, 1999).

and Hispanic students of all ages remained unchanged. In 1999, boys outperformed girls in science at ages 13 and 17, but not at age 9. Among 17-year-olds, the score gap between boys and girls has narrowed since 1969 (Campbell, Hombo, and Mazzeo, 2000).

Like the long-term trend NAEP, the NAEP science achievement test focuses primarily on mastery of subject matter. Between 1996 and 2000, the average score of 12th grade students on this test declined from 154 to 150 (a small but statistically significant amount), while the scores of 4th grade and 8th grade students remained unchanged (National Center for Education Statistics, 2001a). Student performance on the NAEP science achievement test also varied by race and by students' socioeconomic status. In both 1996 and 2000, the average score for white students was higher than the average for black and Hispanic students. Students of lower socioeconomic status, as indicated by their eligibility for free or reduced-price meals, had lower average NAEP science scores than students from more wealthy families (National Center for Education Statistics, 2001b).

Results of International Comparative Tests

Results of international comparisons provide additional insight into the science knowledge of U.S. high school students. TIMSS assessed the science performance of 8th graders in the United States and many other countries in 1995, 1999, and 2003. Over that time period, U.S. 8th graders improved their average science performance slightly, both in comparison with the earlier cohorts of 8th graders and relative to the 44 other countries that participated in the studies (Gonzales et al., 2004). The average scale score in science increased from 513 in 1995 to 527 in 2003, placing the United States well above the international average of 473 among all 8th graders in all participating nations.

Like the framework of the NAEP science achievement test, the TIMSS framework includes both a range of science subject matter and also student abilities related to scientific inquiry and investigations. However, with fewer performance tasks than the NAEP science achievement test, TIMSS may be more limited in its capacity to measure student attainment of the other goals of laboratory experience, besides mastery of subject matter (Owen, 2005).

Results from another international comparative test, PISA, suggest U.S. high school students have not increased their science achievement. In 2000 and 2003, 15-year-old students in many countries took two-hour PISA tests that focused primarily on reading (in 2000) and mathematics (in 2003) and also included some items related to scientific literacy. The U.S. scientific literacy score was below the average among OECD countries in 2000 and in 2003, and there was no measurable change in U.S. students' scores between the two years (Lemke et al., 2004). The PISA science test framework includes several elements that are aligned with the goals of laboratory experiences, including knowledge of science concepts and the ability to apply this knowledge to describe, explain, and predict scientific phenomena; to understand scientific investigations; and to interpret scientific evidence and conclusions. About half of the test items asked students to perform tasks that reflected applications of scientific knowledge to life and health, the environment, and technology, while the other half were selected-response items (Organisation for Economic Co-Operation and Development, 2004).

Overall, then, results from large-scale national and international tests indicate that U.S. high school students have made little or no progress in mastery of science subject matter. Such mastery might be attained through laboratory experiences or through other forms of science instruction, including reading, lectures, discussion, and work with computers. The tests yield little information about the extent to which U.S. high school students may have attained other educational goals of laboratory experiences.

High School Science and Undergraduate Science Achievement

Policies aimed at improving science education are designed in part to prepare more U.S. high school students to enter higher education in science and engineering degrees, in preparation for careers in these fields. The U.S. science and technology workforce is aging, and global competition for skilled scientists and engineers is growing (National Science Foundation, 2004).

Many undergraduate science and engineering students do not complete their degrees. Among first-year students who declared majors in science and engineering in 1990, fewer than half had completed such a degree within five years. Among those who did not complete such a degree, approximately 20 percent of the students dropped out of college, and the remainder chose other fields of study (Huang, Taddese, and Walter, 2000).

Although students drop out of scientific and technology majors for a variety of complex, individual reasons, one important reason may be that their high school science education, including their laboratory experiences, did not adequately prepare them for undergraduate education. A survey conducted in 2002 indicated that 20 percent of first-year students planning to major in science and engineering fields needed remediation in mathematics, and nearly 10 percent reported needing remediation in the sciences (National Science Foundation, 2004). In a recent study of student scores from its widely used college admissions test, the American College Testing Service found that only 26 percent of students tested in 2003-2004 were ready to pass their first college biology course with a grade of "C" or better (American College Testing Service, 2004).

Little research is available on the role that laboratory experiences may play in preparing students to succeed in undergraduate science education. However, one study is available (Sadler and Tai, 2001). The authors surveyed nearly 2,000 undergraduate physics students at public and private institutions and compared their undergraduate physics grades with their high school physics experiences. The analysis of the survey findings indicates that, when demographic factors were controlled, taking a high school physics course had a modest positive effect on undergraduate physics grades.

The researchers also found that students who took high school physics courses that spent more time addressing fewer topics in depth (including fewer concepts, topics, and laboratory activities) had higher undergraduate physics grades than students whose high school physics courses covered more topics in less depth. The authors suggest that high school physics teachers should concentrate on a limited set of topics related to mechanics and include laboratory experiences carefully chosen to reflect those topics. They note, "Doing fewer lab experiments can be very effective if those performed relate to critical issues and students have the time to pursue them fully" (Sadler and Tai, 2001, p. 126). These findings suggest that laboratory experiences may be more effective in supporting student learning when they are integrated into the stream of science instruction, as we discuss further in Chapter 3.

Rising Enrollments and Increasing Diversity

Trends in public understanding of science and science achievement are influenced by larger changes in the U.S. education system. Rising immigration—the total immigrant population of the U.S. nearly tripled from 1970 to 2000—and the baby boom echo—the 25 percent increase in the number of annual births that began in the mid-1970s and peaked in 1990—are boosting school enrollment. After declining during the 1970s and early 1980s, enrollment in public schools increased in the latter part of the 1980s and the 1990s, reaching an estimated 48.0 million in 2003 (National Center for Education Statistics, 2004d).

With rising enrollments, some science teachers face large classes. In California, total statewide enrollment in kindergarten through 12th grade grew from 5.2 million in 1992-1993 to 6.3 million in 2003-2004 (California Education Data Partnership, 2005). In recent years, the average size of science classes grew from 29.3 students in 2000-2001 to 30.1 students in 2003-2004 (California Education Data Partnership 2005).[2]

---

[2]California and other states also report pupil-teacher ratios. This ratio is different from average class size because it is the number of pupils per full-time-equivalent teacher, including teachers who are not in the regular classroom. The pupil-teacher ratio in California high schools declined slightly from 24.5 in 1992-1993 to 23.5 in 2003-2004 (California Education Data Partnership, 2005).

Linguistic and Ethnic Diversity

In concert with these growing enrollments, student diversity has increased. The total proportion of public school students considered to be part of a minority group increased from 17 percent in 1972 to 39 percent in 2000, largely due to rapid growth in the proportion of Hispanic students (National Center for Education Statistics, 2004b). At the same time, poor and minority students are increasingly concentrated in high-poverty schools.

Anthropologists have suggested that groups who are underrepresented in the scientific and technology professions constitute a culture that is different from the culture prevailing in school and in the scientific community (Costa, 1995). Such students cross cultural borders from the world of their peers and family into the world of school science, and conflicts between these different cultures may detract from their learning of science (Cobern and Aikenhead, 1998). They bring everyday knowledge and ways of thinking and talking developed in their home cultures that are rarely acknowledged or used in school (Heath, 1989; Lee, 2000). Researchers have found that teachers who focus on identifying this everyday knowledge can tap it in ways that support students in developing understanding of science concepts (Warren et al., 2001).

A recent review of the research on science education and student diversity concluded that diverse science students may benefit from special support in learning and using scientific language, in becoming comfortable with the community of school science learners, in understanding scientific concepts and modes of thinking, and in developing trusting relationships with other students and the teacher (Lee and Luykx, in press). The authors suggest that laboratory experiences may be particularly valuable in helping the many children of immigrants who are not proficient in English develop improved understanding of science. Students' direct interactions with natural phenomena often require less formal academic language than does reading a textbook or other forms of science instruction. In addition, small-group laboratory experiences can provide structured opportunities for developing language proficiency in a more comfortable environment than speaking in front of the whole class (Lee and Luykx, in press).

Special Educational Needs

In addition to being more racially and linguistically diverse than previous generations of students, today's students also vary more widely in terms of special educational needs. The fraction of students served by federally funded programs for children with disabilities rose from 8.3 percent in 1976-1977 to 13.4 percent in 2001-2002 (U.S. Department of Education, 2003). Some of the rise since 1976-1977 may be attributed to the increasing proportion of students identified as learning disabled, whose share of those with disabilities increased from 21 percent in 1976-1977 to 44 percent in 2001-2002. In 2001-2002, most students with disabilities were those with learning disabilities (44 percent), speech or language impairments (17 percent), mental retardation (9 percent), and emotional disturbance (7 percent). Smaller percentages of students received services for visual, hearing, orthopedic, mobility, or other disabilities.

Mainstreaming of these special needs students has increased in response to federal law. The Individuals with Disabilities Education Act (P.L. 105-17, most recently amended on December 3, 2004), other federal and state laws, and a substantial body of case law give students with disabilities the right to a free and appropriate public education (National Research Council, 1997). The law requires that this education be tailored to individual learning needs, and that each student have an individualized education program stating educational objectives and identifying strategies to attain those objectives. In compliance with the legal requirement to educate disabled students in the "least restrictive environment," more disabled students are in regular classrooms. Between 1988-1989 and 1999-2000, the percentage of students with disabilities spending at least 80 percent of their time in a regular education classroom increased from 31 to 47 percent (National Center for Education Statistics, 2002).

The provisions of the Individuals with Disabilities Education Act requiring that students be placed in the least restrictive setting apply to science laboratory facilities and experiences. In order to provide disabled students with access to laboratory experiences, schools may provide accommodations in laboratory instruction, in the physical design of the laboratory or classroom, or in the ways in which students demonstrate their knowledge of science (Keller, 2002). The level of student involvement in various laboratory activities and the types of accommodations required are often best determined in discussions between the individual teacher and student (Center for Rehabilitation Technology and IMAGINE Group, 2004).

For example, students with learning disabilities may be provided with the course syllabus in advance, may receive extended time for completion of laboratory reports, or may be seated close to the teacher. A student with hearing impairment may be helped by the use of visual aids, while accommodations for students with visual impairment include large lettering, a magnifying glass, and a large notebook (Center for Rehabilitation Technology and IMAGINE Group, 2004). Laboratory experiments modified for students with disabilities are available online (Center for Rehabilitation Technology and IMAGINE Group, 2004), and other resources are available to accommodate the needs of students with disabilities (see for example, Turner, 2004; Miner, Swanson, and Woods, 2001).

Although research on effective approaches to science education for diverse students—including those with limited English proficiency, minority students, low-achieving science students, and students with learning disabilities—is growing, meeting the needs of individual students remains a great challenge for teachers and schools. Chapter 3 discusses promising approaches to laboratory instruction that appear to enhance learning among all students, including those with limited English proficiency, minorities, and low-achieving students.

POLICIES INFLUENCING HIGH SCHOOL LABORATORY EXPERIENCES

Over the past 20 years, the states, the federal government, school districts, and the scientific community have launched an array of efforts to improve science education that may influence high school laboratory experiences. State education policies, including requirements for high school graduation and college admission, science standards, and assessments may affect laboratory instruction. Scientific professional associations, agencies, and research institutions have also developed science education programs and policies, some of which focus specifically on high school laboratories.

State High School Graduation Requirements

High school graduation requirements are one "policy driver" influencing the extent to which high school students enroll in science courses and participate in science laboratory experiences. Between 1982 and 2000, most states increased the number of science courses required for graduation; in response, a growing percentage of high school students completed science courses beyond general biology (National Center for Education Statistics, 2004a). Because most high school science teachers engage students in laboratory experiences at least once per week (Smith et al., 2002), the trend toward taking more science courses translates to an increase in the amount of time the average high school student spends in laboratory experiences (see Chapter 4).

Some states specifically require students to complete laboratory science courses in order to graduate. In 2004, 13 states explicitly mentioned enrollment in a laboratory science course as part of the regular high school graduation requirement (Table 2-1). Of these, five states—Florida, Indiana, New York, South Dakota, and Virginia—required more than one laboratory science course. In addition to these 13 states, 3 states (Arkansas, Kentucky, and Rhode Island) required labs only for an optional college preparatory curriculum or advanced diploma.

TABLE 2-1 State Science Laboratory Requirements for High School Graduation in 2004

State	Laboratory Requirement
Arkansas	College prep only—3 lab courses
District of Columbia	1 lab course
Florida	2 lab courses
Idaho	1 lab course
Indiana	2 science courses (state standards indicate all science courses are to include laboratory activities)
Kansas	1 lab course
Kentucky	College prep only—1 lab course
Maine	1 lab course
Maryland	1 lab course
New Mexico	1 lab course
New York	2 lab courses
Pennsylvania	1 lab course
Rhode Island	College prep only—1 lab course
South Dakota	2 lab courses
Virginia	3 lab course (4 lab courses for college prep)
Washington	1 lab course

SOURCE: Compiled from Sommerville and Yi (2002); Council of Chief State School Officers (2002); state web sites.

State Requirements for Higher Education Admissions

College and university entrance requirements influence the high school curriculum in general and may also influence individual students' decisions about enrolling in science courses, including laboratory science courses. Public and private colleges and universities have varying entrance requirements, but many states have established somewhat uniform standards for entrance into state-supported institutions of higher education. In 2002, 30 states had established the minimum number of courses that students must complete in each discipline to gain admission to public four-year institutions, and 29 required students to have completed at least 2 years of science. Among these 29 state higher education systems, 21 required that at least one of the science courses be a laboratory course (Sommerville and Yi, 2002). Among the 21 states that did not require any specific science courses for admission to higher education, many did require a high school diploma, with its attendant requirements for science courses, sometimes including laboratory courses.

Over the past three decades, the number of high school graduates going directly on to higher education has grown. By 2001, an average of 62 percent of all high school graduates entered colleges or universities (National

Center for Education Statistics, 2004b). In light of these increases, it is interesting to compare state science requirements for high school graduation with state requirements for higher education admission. Data from a study of requirements as of 2002 revealed a mismatch in the number of states requiring laboratory courses for high school graduation (9) and the number of states requiring laboratory courses for college entrance (20).[3] Five states (Florida, Idaho, Maine, Maryland, and Washington) were matched in that at least one science laboratory course was required at both levels. But even within this group, only two states (Florida and Idaho) were perfectly matched in the number of required laboratory courses, while in the remaining three states (Maine, Maryland, and Washington) high school graduation required one laboratory course while college entrance required two. In the four states without this match (Kansas, New Mexico, New York, and Virginia), a laboratory course was required at the high school but not at the college level.

Notably, most states that require a laboratory course for high school graduation or college entrance do not, within those requirements, define what constitutes a laboratory science course. This lack of definitions is one reflection of the larger issue discussed in Chapter 1: researchers and educators do not agree on how to define high school science laboratories or on their purposes in the high school science curriculum.

Science Standards and Assessments

State education policies often focus on identifying clear and specific science standards and creating assessments to measure student attainment of those standards in order to guide improvements in science teaching and learning. However, the goals embodied in state science standards and the ways in which those standards are implemented and assessed do not reflect the full range of educational goals that laboratory experiences may help students attain. These goals include:[4]

- Enhancing mastery of subject matter.
- Developing scientific reasoning.
- Understanding the complexity and ambiguity of empirical work.
- Developing practical skills.
- Understanding the nature of science.
- Cultivating interest in science and interest in learning science.
- Developing teamwork abilities.

---

[3]The lack of alignment between high school graduation requirements and college entrance is apparent in other content areas as well and has been noted in other studies (The Education Trust, 1999).

[4]In Chapter 3, we discuss each goal in greater detail.

Current state science standards and assessments are derived in part from the *National Science Education Standards (NSES)* (National Research Council, 1996). The standards for grades 9-12 include seven elements, several of which are quite similar to the goals of laboratory experiences identified by the committee: (1) science as inquiry, including abilities to conduct scientific inquiry and understandings about scientific inquiry; (2) physical science; (3) life science; (4) earth and space science; (5) science and technology; (6) science in personal and social perspectives; and (7) history and nature of science.

By 2003, most states had adopted science education standards and curriculum frameworks derived at least in part from the NSES (National Research Council, 1996) and the American Association for the Advancement of Science (AAAS) benchmarks (American Association for the Advancement of Science, 1993). In the No Child Left Behind Act of 2002, the federal government strengthened—and added requirements to—existing state educational standards and assessment systems. Among other provisions, the law requires states to administer assessments of science achievement beginning in school year 2007-2008. The law requires that states assess science achievement once each year in each of three grade bands. In order to comply with this federal law, as well as to guide schools and teachers in implementing state science standards, many states have begun to develop and administer annual assessments of students' science learning.

State Science Standards and the Goals of Laboratories

Throughout the history of U.S. science education, educators and scientists have debated the relative importance of exposing students to many science subjects versus engaging them in deeper study of fewer subjects or concepts. In recent years, state science standards have embodied the former approach, including a broad range of science topics (Duschl, 2004; Massel, Kirst, and Hoppe, 1997). In addition to listing topics, many state standards also call for students to engage in laboratory experiences and to develop understanding of processes of scientific investigation. In theory, state standards could be used as flexible frameworks, guiding integration of laboratory experiences with the teaching of science concepts, in order to progress toward all of the science learning goals identified by the committee. In reality, this rarely happens. Instead, state and local officials and science teachers often see state standards as requiring them to help students master the specific science topics outlined for a grade level or science course. When they view laboratory experiences as isolated events that do not contribute to that mastery of subject matter, and science class time is limited, they may devote little class time to laboratory experiences.

The lists of science topics included in state science standards, when viewed in this way, can conflict with other elements of state science standards that call for students to engage in laboratory experiences. For example, California state science standards for high school students include standards for investigation and experimentation (California State Board of Education, 2004). Modeled on the *NSES* inquiry standards for grades 9-12, the California standards call for students to develop questions and perform investigations, select and use appropriate tools, identify and communicate sources of error, identify possible reasons for inconsistent results, formulate explanations, solve scientific problems, distinguish between hypothesis and theory, and achieve other goals related to laboratory learning.

However, California state standards also require students to learn about many science topics, limiting the time they have available to engage in laboratory experiences that might help them attain the investigation and experimentation standards. When one school district official added up all the science topics to be covered in grades 8 through 10 and divided them by the number of school days, she found that the teachers would have only three days to introduce chemistry students to the methods used in calculating the quantities of reactants and products in a chemical reaction (Linn, 2004).

State Science Assessments and the Goals of Laboratories

Current state science assessments are not well suited to assessing student attainment of the goals of laboratory experiences for two reasons. First, state assessments are not always fully aligned with state science standards (Lawrenz and Huffmann, 2002; Webb et al., 2001). Specifically state science assessments are not always aligned with those elements of state standards that call for laboratory experiences and for attainment of laboratory learning goals during the high school years.

Second, current state assessments emphasize mastery of a broad spectrum of science topics and do not measure progress toward such other goals as developing scientific reasoning, understanding the complexity and ambiguity of empirical work, and developing practical skills. Many are primarily composed of selected-response (multiple-choice) tasks. Such assessments can test student knowledge of many items in a relatively short time, can be scored by computer, are relatively inexpensive, and provide a reliable (or consistent) view of student knowledge (National Research Council, 2002).

Although they are well suited to measuring mastery of science subject matter, current state science assessments may not be appropriate for measuring student attainment of the other goals of laboratories (e.g., scientific reasoning, understanding the complexity and ambiguity of empirical work, and understanding of the nature of science). A recent study of three science exams, which are used widely in many states and consist entirely of selected response items, revealed uneven, scant coverage of most of the goals related to student understanding of the processes of science in the NSES (Quellmalz and Kreikemeier, 2004).

Although performance assessments may be used as a supplement to selected-response items in state science assessments, they present new challenges. Generally, performance assessments require test takers to demonstrate their skills or content knowledge in settings that resemble real-life settings. In comparison to assessments composed of many selected-response items, performance assessments present students with fewer, more realistic tasks. But with fewer tasks, performance assessments are often less reliable or consistent in measuring students' science achievement. In addition, because they generally require more time to develop, test, and administer and because they must be scored by humans using detailed scoring rubrics, performance assessments are generally more subjective and more expensive than selected-response tests (Mislevy and Knowles, 2002). Research shows that student scores on traditional selected-response assessments have little correlation with student scores on performance assessments (Shavelson and Ruiz-Primo 1999).

Current science performance assessments have been influenced by earlier generations of hands-on laboratory practical examinations (Duschl, 2004). They can be delivered by pencil and paper exams, by computer, or in a hands-on laboratory format. Pencil and paper tests may include tasks that ask students to explain how they plan and conduct experiments, gather and organize data, interpret data, and communicate results and conclusions (Quellmalz and Moody, 2004).

The experience of the NAEP science achievement test illustrates some of the challenges of using performance tasks to measure the full range of goals of laboratory experiences. The test framework calls for measuring students' conceptual understanding, scientific investigation abilities, and practical reasoning in the fields of earth, physical, and life science (National Center for Education Statistics, 2004c). Within scientific investigation abilities, the framework calls for assessing students' ability to acquire new information, plan scientific investigations, use scientific tools, and communicate results of investigations. About 60 percent of the test items are performance tasks, and 40 percent require a selected response. In 1996, all students conducted a single hands-on task using a uniform kit of science materials to perform an investigation, make observations, record and evaluate experimental results, and apply problem solving skills. However, distributing the kits and training experts to score the hands-on task was a logistical challenge, and in 2000 and 2005, only half of the students in each school conducted a hands-on task.

Although its framework and inclusion of a performance task would suggest that the NAEP science achievement test is capable of measuring student attainment of the goals of laboratory experiences, the test focuses mostly on mastery of science subject matter. A National Research Council (NRC) committee concluded that the 1996 test items and tasks and the accompanying scoring rubrics failed to capture the more complex aspects of the framework and noted that "technology for using performance-type measures in science via the current large-scale survey assessment clearly has serious shortcomings" (National Research Council, 1999, p. 133). Because of such challenges, few states have implemented hands-on performance tasks as part of their state science assessments (see Box 2-1). The states are, however, forming consortia to share expertise and costs as they develop science achievement tests in response to the mandate of the No Child Left Behind Act and consider the possibilities for including written, hands-on, or computerized performance tasks. These consortia may draw on online collections of performance tasks and other data banks of science test items (Quellmalz and Moody, 2004). Guidance is also available from a recent NRC study of test design for science achievement (National Research Council, 2005).

Implementing State Standards

Although state science standards often embody goals related to mastery of subject matter, they sometimes include at least some of the other goals of laboratory experiences, and some state science standards specifically call for students to participate in laboratory investigations. Studies of local implementation of state science standards indicate that these policies primarily affect coverage of science content and have less influence on teaching methods, including decisions about when and how to include laboratory instruction.

The extent to which the goals of state science standards, including the goals related to laboratory experiences, are implemented depends on local agents and agencies. One important agency is the local school district. Numerous studies, in states from Maine to California, suggest that district policy makers, teachers, and school administrators not only heed state policies but also work hard to implement them (Educational Evaluation and Policy Analysis, 1990; Finnigan and Gross, 2001; Firestone, Fitz, and Broadfoot, 1999; Hill, 2001). A study of the local response to state mathematics and science standards in Michigan in the mid-1990s concluded that school district policy makers and teachers paid close attention to state policy, especially the assessment component and the sanctions that state policy makers had attached to them (Spillane, 2004). According to district policy makers, state sanctions were especially influential in motivating them to develop or revise their instructional policies. Other studies, in Maryland, Washington, and Chicago revealed similar patterns of close attention to state standards and accountability systems (Koretz et al., 1996; Lane et al., 2000; Stecher et al., 2000; Finnigan and Gross, 2001; Kelly et al., 2000).

BOX 2-1 **Hands-On Performance Assessment of Laboratory Learning: The Experiences of New York and Vermont**

New York. In recent years, New York has increased the number of science courses required for high school graduation. The state now requires all high school students to complete three science courses, including one Regents science course that incorporates 1,200 minutes of laboratory activity in order to graduate (Champagne and Shiland, 2004). To assess laboratory learning in these Regents science courses, the state has for many years administered a Regents Examination in each subject consisting of both a written test with tasks related to laboratories (such as items asking about laboratory techniques and design of experiments) and a laboratory performance test. However, state science teachers and education officials grew concerned about the validity and reliability of the performance tests and their alignment with the state laboratory science standards.

In 2002, state officials convened four design teams to develop new performance assessments of laboratory learning as part of the Regents examinations in earth science, chemistry, biology, and physics. The planned new Physical Setting/Earth Science Performance Test included hands-on tasks to be completed at six stations in a secure laboratory classroom. Students were to be tested on their ability to identify minerals, locate an earthquake epicenter, measure atmospheric moisture, determine the density of different fluids, collect and analyze data on the settling of particles in a column of fluid, and construct and analyze an elliptical orbit (DeMauro, 2002).

Initial plans called for introducing the new Physical Setting/Earth Science Performance Test in June 2004. The planned tests posed a challenge to schools and teachers with their requirements for dedicated laboratory testing space and scheduling of students and test administrators. In addition, further research was needed to determine the validity and reliability of the test and to ensure security of test items (Champagne and Shiland, 2004). Because of these challenges, implementation of the new tests has been postponed until 2007.

Vermont. Vermont has chosen a slightly different approach to performance assessment of laboratory learning. The state joined the Partnership for the Assessment of Standards-Based Science (PASS) in 2000. Funded by the National Science Foundation, the PASS is an assessment system designed to allow states and districts to measure students' scientific literacy, as defined by the AAAS Benchmarks (American Association for the Advancement of Science, 1993) and in the *National Science Education Standards* (National Research Council, 1996). The PASS assessment includes four components (WestEd, 2004):

- Hands-on performance tasks,
- Constructed-response investigations,
- Open-ended questions, and
- Enhanced multiple-choice questions.

A 1999 content analysis by a group of scientists and teachers identified a close alignment between Vermont state science standards and the *NSES*. Following the decision to join PASS, the state and WestEd worked together to modify the test to ensure it would accurately measure attainment of Vermont state standards. The modified test for grades 9 and 11 was designed to measure specific aspects of the national standards for science inquiry.

After administration of the PASS test began in 2000, teachers sought more information about how to design and implement laboratory learning (Carvallas, 2004).

Because the performance assessment components of the PASS assessment were administered in regular classrooms using kits of hands-on materials, the test presented fewer logistical challenges and did not include the costs of providing secure laboratory classroom space that was required in New York. However, the test was expensive, time-consuming, and difficult to administer, and it was discontinued after the 2002-2003 school year. Vermont is currently joining forces with two neighboring states (New Hampshire and Rhode Island) to seek economies of scale in performance assessment (Pinckney and Taylor, 2004).

However, a number of studies suggest that the response to state policy at the school district and classroom levels often involves surface changes focusing on content coverage (mastery of subject matter) rather than the broader and more substantive shifts called for in the NSES and in some state standards (Spillane, 2004; Firestone et al., 1996; Spillane and Zeuli, 1999). This is a particularly serious concern with regard to high school science laboratories, because, as we discuss in Chapter 5, using laboratory experiences to advance the science learning goals identified by the committee requires deep and substantial shifts in teaching strategies.

Some research indicates that school districts respond to state standards by focusing primarily on only one of the goals of laboratory experiences—enhancing mastery of science subject matter. For example, the analysis of nine Michigan school districts found that district policies provided strong and consistent support for state policy with respect to coverage and sequencing of topics. District policies' support for other aspects of the state's mathematics and science standards was not nearly as prominent or as faithful as their support for topic coverage and sequencing (Spillane, 2004).

Despite considerable effort by district officials, district policies in six of the Michigan districts provided relatively weak or low support for the mathematics and science standards. Only four districts, for example, provided strong or high support for the more complex changes in mathematics and science education advanced by standards, such as the types of changes required to help students attain the full range of the goals of laboratory experiences. These patterns were repeated at the classroom level in the nine school districts; teachers attempting to implement the reform taught in ways that diverged fundamentally from the intent of the designers (Spillane and Zeuli, 1999).

In a study of teachers' responses to state policy in Maine and Maryland, Firestone and his colleagues found similar patterns with state policy having considerable success in aligning what subjects were taught but less success in changing instructional strategies (Firestone et al., 1999). Similarly, in a study of Kentucky and North Carolina, McDonnell and Chossier (1997) found that while teachers adopted new teaching strategies in response to state policy, the depth of their content and teaching did not change in meaningful ways.

This growing body of research on local implementation of state science and mathematics standards indicates that state standards appear to affect teachers' decisions about coverage, while having less influence on instructional strategies (Hill, 2001; Spillane and Zeuli, 1999). Instructional strategies include the particular materials and methods (which may include laboratory experiences, creating instructional groups, lecturing, leading discussions, etc.) that teachers use to engage students with subject matter.

A recent observational study of a national sample of K-12 mathematics and science lessons involving 364 teachers in 31 schools provides further evidence of the limited impact of state science standards on teachers' instructional strategies (Weiss et al., 2003). Trained observers rated science lessons in terms of design, implementation, content addressed, and classroom culture, and they also conducted detailed interviews with teachers about factors that may have influenced the content and methods used in the observed lesson. The teachers indicated that state and district curriculum standards were especially influential in determining the content covered, influencing more than 7 of 10 lessons observed nationally. With respect to teaching strategies, however, these policy documents were much less influential: only 5 percent of the teachers in the study reported that state or district curriculum standards or frameworks were influential. Teachers reported having a great deal of autonomy in choosing teaching strategies: in 9 of 10 lessons observed, the teacher indicated his or her own knowledge, beliefs, and experiences as the most salient influence.

Effective laboratory teaching that helps students master subject matter, develop scientific reasoning, and progress toward the other goals we identify requires not only deep knowledge of science content but also pedagogical knowledge. The research outlined above suggests that current state science standards do not yet successfully support the deeper changes in teaching strategy necessary to help students attain the educational goals of laboratory experiences.

The uneven local response to state science standards is in part a problem of uneven support for teachers rather than local resistance to change. State standards that press complex changes departing radically from extant practice—such as those calling for laboratory experiences and for attainment of a range of laboratory learning goals—are unlikely to succeed in changing classroom practice unless teachers are supported in developing new understandings about science, teaching, and learning (see Chapter 5).

The Influence of Curriculum on Science Instruction

In contrast to state science standards, science texts and curriculum packages appear to have a greater impact on teaching methods. In about 7 of 10 lessons observed, the teachers interviewed in the observational study discussed above said that textbooks or curricular programs (or both) had influenced their teaching strategies (Weiss et al., 2003). In response to a larger national survey conducted in 2000, science teachers indicated that, in 95 percent of their most recent science classes, they had used commercially published textbooks and related materials (Smith et al., 2002). Since textbooks, along with teachers' own knowledge and beliefs, strongly influence their instruction, the way these texts treat laboratory experiences appears important.

The TIMSS conducted in 1995 provides information about how laboratory experiences are treated in textbooks and state curriculum frameworks. As part of TIMSS, experts gathered information about curriculum standards and textbooks in almost 50 countries for 13-year-old students (8th graders in the United States) and for students in their final year of secondary education. The final-year students included "generalists," those who were in vocational programs, and "specialists," those taking advanced courses in physics.

The study sampled curriculum guides and textbooks in the participating countries. Curriculum guides were defined as official documents that most clearly reflected the intentions, visions, and aims of curriculum makers. In the United States, where national guides are not available, state guides were analyzed.[5] Importantly, any lab manuals provided with a textbook were included in the analysis. Guides and textbooks were selected to represent those in use with at least half of the students in the targeted grade.

The TIMSS researchers developed a common framework to analyze the science curriculum materials across all countries. The framework included performance expectations, some of which are clearly relevant to laboratory experiences:

- theorizing, analyzing, and solving problems;
- using tools, routine procedures, and science processes; and
- investigating the natural world.

The researchers found that, although U.S. state curriculum guides for 8th grade science education referred to each of the three expectations, less than 10 percent of textbook content was devoted to helping students develop these laboratory-related abilities. They found textbooks in most other countries studied devoted a similarly low level of attention to these performance expectations, except for Germany, Hong Kong, and New Zealand, where coverage was slightly greater (21-40 percent of textbook content).

For 12th grade students taking advanced physics in the United States, no information was available from state curriculum guides, and only 10 percent of the content of textbooks addressed these three performance expectations related to laboratory experiences. This degree of coverage was similar to that in other countries. These results suggest that textbooks and the materials accompanying them give little attention to the learning goals of laboratory experiences, even though they may be identified as a priority in state science standards and curriculum guides.

The lack of attention to laboratory experiences in curriculum guides and textbooks may reflect state policies emphasizing coverage of a broad range

---

[5]The study examined curriculum guides used by the states in 1992-1993, prior to release of the *National Science Education Standards*.

of science content. TIMSS found that grade 8 textbooks in the United States covered 65 science topics compared with about 25 topics typical of other TIMSS countries. The authors note that (Valverde and Schmidt, 1997, p. 3): "U.S. eighth-grade science textbooks were 700 or more pages long, hardbound, and resembled encyclopedia volumes. By contrast, many other countries' textbooks were paperbacks with less than 200 pages."

Another study, focusing on high school biology texts, indicated that the most widely used texts provided little support for student learning through laboratory experiences. AAAS developed and applied a detailed protocol to 10 widely used biology curricula, including 4 developed with NSF support. AAAS found that all of these curricula (which included kits of laboratory materials) did a poor job in terms of two criteria that might reflect laboratory experiences: (1) engaging students with relevant phenomena and (2) helping them to develop and use scientific ideas (American Association for the Advancement of Science, 2000).

A panel convened by NSF to review its middle school science curricula gave the texts generally high marks (3 or higher on a 5-point scale) and found that they were consistent with the *NSES*. However, the panel noted a lack of attention to one of the goals of laboratory experiences—enhanced understanding of the nature of science—in these curricula (National Science Foundation, 1997). In a subsequent review of a sample of NSF-funded curricula for elementary, middle, and secondary mathematics and science, experts gave the science curricula high marks (on a 1 to 5 scale) on several criteria that reflect the goals of laboratory experiences, including:

- Do the materials provide sufficient activities for students to develop a good understanding of key science concepts? **4.5**
- Do the materials accurately represent views of science as inquiry? **4.4**
- To what extent do the materials provide students the opportunity to make conjectures, gather evidence, and develop arguments to support, reject, and revise their preconceptions and explanations? **4.3**

In this evaluation, the panel found that, although the content of the curriculum materials (including laboratory kits) was generally high, dissemination was limited. Often, curriculum materials were adopted by a single teacher, rather than a school or a school district. Most large textbook publishers chose not to develop commercial versions of NSF-funded materials (Tushnet et al., 2000). The panel also found that teachers and administrators were unaware of the full range of materials available, and teachers were often unprepared for the changes in instructional strategy required to successfully implement the curricula.

In conclusion, curriculum materials appear to influence science teachers' teaching strategies, including decisions about when and how to engage students in laboratory experiences. The limited evidence available suggests that some curriculum materials are available to support teachers and students in effective laboratory experiences, but these materials are not widely used. The most widely used science texts and accompanying laboratory materials do not reflect the science learning goals of laboratory experiences. Involving teachers in the design, selection, and implementation of curriculum materials and providing professional development aligned with those materials appear essential for successful implementation (Tushnet et al., 2000).

The Role of the Scientific Community

Policies and programs initiated by the scientific community may also influence high school laboratory experiences. NSF evaluates research proposals and provides funding based not only on intellectual merit but also on "broader impact" (including impact on education), and the National Aeronautics and Space Administration (NASA) requires that a small percentage of funds for each large space mission be set aside for public outreach, including education. The National Institutes of Health (NIH) and the Department of Energy also support programs aimed at improving high school science education. Many scientific societies, including the American Chemical Society, the American Physiological Society, and the American Institute of Biological Sciences, are also working to improve science education. Congress provides a stream of funding for partnerships between scientists and educators through the Math-Science Partnerships programs, and private agencies, such as the Howard Hughes Medical Institute (HHMI), also support efforts to improve school science education. In addition, many individual scientists, companies, universities, teachers, and schools are working together to improve high school science courses, including laboratory teaching and learning. To date, however, there has been no systematic effort to assess the scope of these diverse activities and their impact on the science achievement of high school students.

The committee identified several types of efforts by the scientific community that may influence high school laboratory experiences, including programs designed to (1) provide laboratory-centered curricula for use in high schools, (2) provide laboratory facilities and equipment to schools, (3) provide research internships to students and teachers, and (4) provide undergraduate education and professional development to prospective and current science teachers. Here we briefly discuss efforts focused on schools and students; the scientific community's role in teacher education is discussed in Chapter 5.

Providing Laboratory-Focused Curriculum

Scientific agencies and professional societies support development and dissemination of high school science curricula. For example, NSF has supported the American Geological Institute, the American Chemical Society, and the American Institute of Physics in developing and disseminating high school science curricula that incorporate laboratory experiences (Biological Sciences Curriculum Study, 2001). Unlike traditional texts that may be accompanied by a separate laboratory manual, these curricula integrate laboratory experiences into the flow of instruction. The American Geological Institute is also producing a series of DVDs for use in schools that encompass the U.S. Geological Survey's Global Geographic Information System database (Smith, 2004). The Association for Biology Laboratory Education publishes an online *Labstracts* newsletter that provides a variety of laboratory exercises (Association for Biology Laboratory Education, 2005). Although many of these laboratory exercises are provided by undergraduate educators, they can be used by high school teachers as well. Other scientific and teaching societies, in each of the science disciplines, are engaged in similar efforts.

Providing Laboratory Facilities and Equipment

One concrete way in which the scientific community can support high school laboratory experiences is through providing laboratory facilities and equipment. A few such efforts are described here.

San Mateo, California, high school teacher Ellyn Daugherty developed the San Mateo Biotechnology Career Pathway program at San Mateo High School with the help of many local biotechnology companies and foundations. Support from biotechnology firms helped in converting a shop classroom into a large, modern biotechnology classroom and in providing necessary equipment and supplies. Currently, 20 industry partners provide internships to advanced high school students enrolled in the program (Daugherty, 2004). These firms often hire graduates of the high school program, either directly after high school or after two to four years of further biotechnology or biology education.

With support from federal, state, and private agencies, scientists at higher education institutions in several states have designed and equipped mobile laboratories to serve students in schools that lack adequate science facilities (see Chapter 6). For example, during the 1999-2000 school year, the chemistry department at Virginia Polytechnic Institute (VPI) developed a mobile chemistry laboratory to help rural teachers and students respond to Virginia's science standards and assessments, called the Standards of Learning (SOL). The chemistry department convened meetings of teachers from rural high schools in Appalachia and southern Virginia to design and evaluate a series of laboratory experiments aligned with the chemistry SOLs. The following summer, a team of VPI staff, including two chemistry teachers, a lab technician, and an administrative assistant, led the first of a continuing series of summer workshops to train teachers on the experiments and instrumentation available on the van. The VPI team also developed kits of chemistry experiments that did not require advanced instrumentation and began mailing them to rural schools.

Chemistry teachers and students at 19 rural Virginia high schools and two inner-city Richmond schools conducted the experiments included in the mobile van four times during the academic year and also received four to six chemistry kits for each of three years, beginning in the 2001-2002 school year. During the summers, more than 63 teachers were trained in leading the experiments included on the van. Before the mobile van program was initiated, students in these 19 schools performed on average 15.6 percent lower than the state average on the chemistry SOL. In 2003, the average among these 19 schools was 1.2 points above the state average, with particularly large gains in two inner-city Richmond schools with large minority populations. Attendance also improved on the days the mobile van was present, but without a comparison group it is not possible to know whether the mobile van or other factors may have accounted for the improvements in test scores and attendance (Long, 2004). However, the Virginia Tech program, which relied on a combination of federal, corporate, private, and university grants—could not be sustained and was ended in the summer of 2004.

The Virginia program was modeled on the Science in Motion program of Juniata College that serves rural schools in Pennsylvania. An independent evaluation of the Juniata program conducted in 1999 found statistically significant gains in biology and chemistry achievement test scores among students served by the program when compared with students in schools without access to the program (Mulfinger, 2004).

Providing Student Internships

Scientists have provided laboratory internships to high school students for many years. One scientist who involved high school students in a 1955 summer program in the departments of biochemistry and physiological chemistry at the University of California, Berkeley, described them as "enthusiastic, hard-working and intelligent laboratory assistants—well worth the time and cost of training" (Pardee, 1956, p. 725). Scientific agencies and foundations, as well as individual science departments, support such programs. For example, the Howard Hughes Medical Institute provides funding for high school students and teachers in the Montgomery County, Maryland, public schools to study and work in laboratories at the NIH alongside some of the world's leading biomedical scientists. Students who participate in this research program present their results at an annual symposium at HHMI headquarters. In addition to hosting these interns, NIH provides supplements to research grants for the purpose of providing internships to underrepresented minorities. The National Human Genome Research Institute of NIH also provides summer internships to high school students. Such internship opportunities are not restricted to large, government laboratory settings such as NIH. The Noble Foundation supports summer research internships in applied agriculture and plant science for high school students in Oklahoma (Noble Foundation, 2005).

Laboratory internships are often designed to encourage disadvantaged or minority high school students to choose science careers. The American Chemical Society's Project SEED provides students with summer research internships guided by scientist-mentors. Students who are eligible and qualified may participate in Summer I internships before their senior year, in Summer II internships in the summer following graduation, and they may receive first-year college scholarships to study chemistry (American Chemical Society, 2005). Studies indicate that having personal contact with a scientist affects students' preference for and persistence in science careers, and that minority students may be especially encouraged to persist in science studies by contact with minority scientists and engineers (Hill, Pettus, and Hedin, 1990; Barton, 2003).

SUMMARY

Most people in this country lack the basic understanding about science that they need to make informed decisions about the many scientific issues affecting their lives. Neither this basic understanding—often referred to as scientific literacy—nor an appreciation for how science has shaped society and culture—is being cultivated during the high school years. For example, over the 30 years between 1969 and 1999, high school students' scores on the science portion of the NAEP remained stagnant.

State policies regarding student laboratory experiences, including graduation requirements, higher education requirements, state science standards, and assessments, do not always support effective laboratory teaching and learning. Although state science standards could be used as flexible frameworks to guide schools and teachers in integrating laboratory experiences with the teaching of science concepts, this rarely happens. Instead, state and local officials and science teachers often see state standards as requiring them to help students master the specific science topics outlined for a grade level or science course. State science standards that are interpreted as encouraging the teaching of extensive lists of science topics in a given grade may discourage teachers from spending the time needed for effective laboratory learning.

Some state science standards call for students to engage in laboratory experiences and to attain other goals of laboratory experiences, such as developing scientific reasoning and understanding the nature of science. However, assessments in these states rarely include items designed to measure student attainment of these goals. Current large-scale assessments are not designed to accurately measure student attainment of all of the goals of laboratory experiences. Developing and implementing improved assessments to encourage effective laboratory teaching would require large investments of funds.

REFERENCES

American Association for the Advancement of Science. (1993). *Benchmarks for science literacy.* Washington, DC: Author.

American Association for the Advancement of Science. (2000). *Big biology books fail to convey big ideas*, Reports AAAS's Project 2061. Washington, DC: Author. Available at: http://www.project2061.org/press//pr000627.htm [accessed Sept. 2004].

American Chemical Society. (2005). *High school-project seed.* Available at: http://www.chemistry.org/portal/a/c/s/1/acsdisplay.html?DOC=education%5Cstudent%5Cprojectseed.html [accessed Jan. 2005].

American College Testing Service (ACT). (2004). *Crisis at the core: Preparing all students for college and work.* Iowa City, IA: Author. Available at: http://www.act.org/path/policy/index.html [accessed Dec. 2004].

Association for Biology Laboratory Education. (2005). *Labstracts: Newsletter of the Association for Biology Laboratory Education.* Available at: http://www.zoo.utoronto.ca/able/news/winter05/index.html [accessed May 2005].

Barton, P.E. (2003). *Hispanics in science and engineering: A matter of assistance and persistence.* Princeton, NJ: Educational Testing Service.

Biological Sciences Curriculum Study. (2001). *Profiles in science: A guide to NSF-funded high school instructional materials.* Colorado Springs, CO: Author.

California Education Data Partnership. (2005). *Ed-data: Fiscal, demographic, and performance data on California's K-12 schools.* Sacramento: Author. Available at: http://www.ed-data.k12.ca.us/ [accessed May 2005].

California State Board of Education. (2004). *Investigation and experimentation-grades 9 to 12: Science content standards.* Sacramento: Author. Available at: http://www.cde.ca.gov/be/st/ss/scinvestigation.asp.

Campbell, J.R., Hombo, C., and Mazzeo, J. (2000). *NAEP 1999—Trends in academic progress: Three decades of student performance: Executive summary.* Washington, DC: U.S. Department of Education, National Center for Education Statistics. Available at: http://www.nces.ed.gov/nationsreportcard//pubs/main1999/2000469.asp [accessed Dec. 2004].

Carvallas, E. (2004). Remarks made to the NRC Teacher Advisory Council meeting, March 3, Washington, DC.

Center for Rehabilitation Technology and IMAGINE Group. (2004). *Barrier free education*. Atlanta: Georgia Institute of Technology. Available at: http://www.barrier-free.arch.gatech.edu/index.html [accessed Dec 2004].

Champagne, A., and Shiland, T. (2004). *Large-scale assessment and the high school science laboratory*. Presentation to the Committee on High School Science Laboratories: Role and Vision. June 4. Available at: http://www7.nationalacademies.org/bose/June_3-4_2004_High_School_Labs_Meeting_Agenda.html [accessed Jan. 2004].

Coburn, W.W., and Aikenhead, G.S. (1998). Cultural aspects of learning science. In B.J. Fraser and K.G. Tobin (Eds.), *International handbook of science education* (pp. 39-52). London, England: Kluwer Academic.

Commission on Instructionally Supportive Assessment. (2001). *Building tests to support instruction and accountability: A guide for policymakers.* Available at: http://www.aasa.org/issues_and_insights/assessment/Building_Tests.pdf [accessed Dec. 2004].

Costa, V.B. (1995). When science is another world: Relationships between worlds of family, friends, school and science. *Science Education, 79*, 313-333.

Council of Chief State School Officers. (2002). *Key state education policies on PK-12 education: 2002*. Washington, DC: Author.

Daugherty, E. (2004). *The San Mateo biotechnology career pathway*. Presentation to the National Research Council Committee on High School Science Laboratories: Role and Vision. July 12. Available at: http://www7.nationalacademies.org/bose/July_12-13_2004_High_School_Labs_Meeting_Agenda.html [accessed Oct. 2004].

DeMauro, G.E. (2002). *Regents examination in physical setting/earth science performance test 2004*. Memorandum. Albany, NY: State Education Department. Available at: http://www.emsc32.nysed.gov/osa/scire/sciearch/regphysearchscimem02.pdf [accessed Jan. 2004].

Duschl, R. (2004). *The HS lab experience: Reconsidering the role of evidence, explanation, and the language of science*. Paper prepared for the Committee on High School Science Laboratories: Role and Vision. Available at: http://www7.nationalacademies.org/bose/July_12-13_2004_High_School_Labs_Meeting_Agenda.html [accessed Dec. 2004].

Educational Evaluation and Policy Analysis. (1990). Educational evaluation and policy analysis. *Quarterly Publication of the American Educational Research Association, 12*(3), 331-338.

Finnigan, K., and Gross, B. (2001). *Teacher motivation and the Chicago probation policy*. Paper presented at the annual meeting of the American Educational Research Association, Seattle.

Firestone, W.A., Fitz, J., and Broadfoot, P. (1999). Power, learning, and legitimation: Assessment implementation across levels in the United States and the United Kingdom. *American Educational Research Journal, 36*(4), 759-796.

Gonzales, P., Guzman, J.C., Partelow, L., Pahlke, E., Jocelyn, L., Kastberg, D., and Williams, T. (2004). *Highlights from the Trends in International Mathematics and Science Study (TIMSS) 2003*. Washington, DC: U.S. Department of Education, National Center for Education Statistics. Available at: http://www.nces.ed.gov/pubsearch/pubsinfo.asp?pubid=2005005 [accessed Dec. 2004].

Heath, S.B. (1989). Oral and literate traditions among black Americans living in poverty. *American Psychologist, 44*(2), 367-373.

Hill, H. (2001). Policy is not enough: Language and the interpretation of state standards. *American Educational Research Journal, 38*(2), 289-318.

Hill, O.W., Pettus, C., and Hedin, B.A. (1990). Three studies of factors affecting the attitudes of blacks and females toward the pursuit of science and science-related careers. *Journal of Research in Science Teaching, 27*(4), 289-314.

Huang, G., Taddese, N., and Walter, E. (2000). *Entry and persistence of women and minorities in college science and engineering education.* (NCES 2000–601.) Washington, DC: National Center for Education Statistics. Available at: http://www.nces.ed.gov/pubsearch/pubsinfo.asp?pubid=2000601 [accessed Dec. 2004].

Keller, E. (2004). *Inclusion in science education for students with disabilities.* Morgantown, WV: Eberly College of Arts and Sciences, West Virginia University. Available at: http://www.as.wvu.edu/~scidis [accessed Dec. 2004].

Kelley, C., Odden, A., Milanowski, A., and Heneman, H. (2000, February). *The motivational effects of school-based performance awards.* CPRE Policy Briefs RB-29. Philadelphia, PA: Consortium for Policy Research in Education. Available at: http://www.cpre.org/Publications/rb29.pdf [accessed Oct. 2005].

Koretz, D., Mitchell, K., Barron, S., and Keith, S. (1996). *Perceived effects of the Maryland State Assessment Program* (Technical Report No. 406). Los Angeles: University of California, National Center for Research on Evaluation, Standards, and Student Testing.

Lane, S., Stone, C., Parke, C., Hansen, M., and Cerrillo, T. (2000). *Consequential evidence for MSPAP from the teachers, principal, and student perspective.* Paper presented at the annual meeting of the National Council of Measurement in Education, New Orleans.

Lawrenz, F., and Huffman, D. (2002). The archipelago approach to mixed methods evaluation. *American Journal of Evaluation, 23,* 331-338.

Lee, C. (2000). *The state of knowledge about the education of African Americans.* Paper prepared for the Commission on Black Education, American Educational Research Association. Available at: http://www.coribe.org.

Lee, O., and Luykx, A. (in press). *Science education and student diversity: Synthesis and research agenda.* New York: Cambridge University Press. Summary available at: http://www.crede.org/synthesis/sd/sd_exsum.html [accessed May 2005].

Lemke, M., Sen, A., Pahlke, E., Partelow, L., Miller, D., Williams, T., Kastberg, D., and Jocelyn, L. (2004). *International outcomes of learning in mathematics literacy and problem solving: PISA 2003 results from the U.S. perspective.* Washington, DC: National Center for Education Statistics (NCES-2005003). Available at: http://www.nces.ed.gov/pubsearch/pubsinfo.asp?pubid=2005003 [accessed Jan. 2005].

Linn, M. (2004). *High school science laboratories: How can technology contribute?* Presentation to the Committee on High School Science Laboratories: Role and Vision. June. Available at: http://www7.nationalacademies.org/bose/June_3-4_2004_High_School_Labs_Meeting_Agenda.html [accessed April 2005].

Long, G. (2004). Mobile chemistry laboratory: Outreach to high schools. In D. Haase and S. Schulze (Eds.), *Proceedings of the Conference on K-12 Outreach from University Science Departments: 2004.* Raleigh: Science House, North Carolina State University.

Massell, D., Kirst, M., and Hoppe, M. (1997). *Persistence and change: Standards-based reform in nine states.* (CPRE Research Report.) Philadelphia, PA: Consortium for Policy Research in Education.

McDonnell, L., and Chossier, C. (1997). *Testing and teaching: Local implementation of new state assessments.* Los Angeles: University of California, National Center for Research on Evaluation, Standards, and Student Testing.

Miller, J.D. (2004). Public understanding of, and attitudes toward, scientific research: What we know and what we need to know. *Public Understanding of Science, 13*(3), 273-294.

Miner, D.L., Nieman, R., Swanson, A., and Woods, M. (2001). *Teaching chemistry to students with disabilities: A manual for high schools, colleges, and graduate programs* (4th edition). Washington, DC: American Chemical Society. Available at: http://www.membership.acs.org/C/CWD/TeachChem4.pdf [accessed Dec. 2004].

Mulfinger, L. (2004). What is good science education, and whose job is it to support it? In D. Haase and S. Schulze (Eds.), *Proceedings of the Conference on K-12 Outreach from University Science Departments: 2004.* Raleigh: Science House, North Carolina State University.

National Center for Education Statistics. (2001a). *National average science scale score results.* Washington, DC: Author. Available at: http://www.nces.ed.gov/nationsreportcard/science/results/natscalescore.asp [accessed Dec. 2004].

National Center for Education Statistics. (2001b). *Science: National subgroup results.* Washington, DC: Author. Available at: http://www.nces.ed.gov/nationsreportcard/science/results/natsubgroups.asp [accessed May 2005].

National Center for Education Statistics. (2002). *The condition of education 2002. (NCES 2002-025).* Washington, DC: Author.

National Center for Education Statistics. (2004a). *Contexts of elementary and secondary education: Coursetaking and standards.* Available at: http://www.nces.ed.gov/programs/coe/2004/section4/indicator21.asp [accessed May 2005].

National Center for Education Statistics. (2004b). *Participation in education.* Available at: http://www.nces.ed.gov/programs/coe/2004/section1/index.asp [accessed May 2005].

National Center for Education Statistics. (2004c). *What does the NAEP science assessment measure?* Washington, DC: Author. Available at: http://www.nces.ed.gov/nationsreportcard/science/whatmeasure.asp [accessed Dec. 2004].

National Center for Education Statistics. (2004d). *Participation in education: Elementary and secondary education.* Available at: http://www.nces.ed.gov/programs/coe/2004/section1/indicator04.asp [accessed May 2005].

National Research Council. (1996). *National science education standards.* National Committee on Science Education Standards and Assessment. Center for Science, Mathematics, and Engineering Education. Washington, DC: National Academy Press.

National Research Council. (1997). *Educating one and all: Students with disabilities and standards-based reform.* Committee on Goals 2000 and the Inclusion of

Students with Disabilities, L.M. McDonnell, M.J. McLaughlin, and P. Morison (Eds.). Washington, DC: National Academy Press.

National Research Council. (1999). *Grading the nation's report card: Evaluating NAEP and transforming the assessment of educational progress.* Committee on the Evaluation of National and State Assessments of Educational Progress, J.W. Pellegrino, L.R. Jones, and K.J. Mitchell (Eds.). Board on Testing and Assessment, Commission on Behavioral and Social Sciences and Education. Washington, DC: National Academy Press.

National Research Council. (2002). *Performance assessments for adult education: Exploring the measurement issues. Report of a workshop.* Committee for the Workshop on Alternatives for Assessing Adult Education and Literacy Programs, Robert J. Mislevy and Kaeli T. Knowles (Eds.). Board on Testing and Assessment, Center for Education, Division of Behavioral and Social Sciences and Education. Washington, DC: National Academy Press.

National Research Council. (2005). *Systems for state science assessment.* Committee on Test Design for K-12 Science Achievement. M.R. Wilson and M.W. Bertenthal (Eds.). Board on Testing and Assessment, Center for Education. Division of Behavioral and Social Sciences and Education. Washington, DC: The National Academies Press.

National Science Foundation. (1997). *Review of instructional materials for middle school science.* Arlington, VA: Author.

National Science Foundation. (2004). *Science and engineering indicators 2004.* Arlington, VA: Author. Available at: http://www.nsf.gov/sbe/srs/seind04/c0/c0s1.htm [accessed Oct. 2004].

Noble Foundation. (2005). *Noble Foundation internship programs.* Available at: http://www.noble.org/Internships/index.html [accessed April 2005].

Organisation for Economic Co-Operation and Development. (2004). *Learning for tomorrow's world: First results from PISA 2003.* Paris: Author. Available at: http://www.pisa.oecd.org/dataoecd/59/21/33917683.pdf [accessed Jan. 2005].

Owen, E. (2005). *Comparing NAEP, TIMSS, and PISA in mathematics and science.* Washington, DC: National Center for Education Statistics, U.S. Department of Education. Available at: http://www.nces.ed.gov/timss/pdf/naep_timss_pisa_comp.pdf [accessed Jan. 2005].

Pardee, A.B. (1956). High school students as laboratory assistants. [Letter to the editor]. *Science Magazine, New Series, 123*(3200), 725.

Pinckney, E.R., and Taylor, G. (2004, July 20). *Memorandum to superintendents of schools, principals, curriculum coordinators, VTSAC re Vermont science assessment update.* Montpelier: Vermont Department of Education.

Quellmalz, E.S., and Kreikemeier, P. (2004). *Testing the alignment of items to the NSES inquiry standards.* Paper presented at the annual meeting of the American Educational Research Association, San Diego.

Quellmalz, E.S., and Moody, M. (2004). *Models for multi-level state science assessment systems.* Paper prepared for the Workshop on Designing High-Quality Science Assessment Systems, National Research Council, May 6-7. Available at: http://www7.nationalacademies.org/bota/Test_Design_K-12_Science.html [accessed April 2005].

Sadler, P., and Tai, R. (2001). Success in introductory college physics: The role of high school preparation. *Science Education, 85*(111), 136.

Shavelson, R.J., and Ruiz-Primo, M.A. (1999). On the psychometrics of assessing science understanding. In J. Mintzes, J.H. Wandersee, and J.D. Novak (Eds.), *Assessing science understanding* (pp. 304-341). Educational Psychology Series. San Diego: Academic Press.

Smith, M. (2004). *High school laboratories: A view from earth and space.* Presentation to the Committee on High School Science Laboratories: Role and Vision. March 29. Available at: http://www7.nationalacademies.org/bose/March_29-30_2004_High_School_Labs_Meeting_Agenda.html [accessed Jan. 2005].

Smith, P.S., Banilower, E.R., McMahon, K.C., and Weiss, I.R. (2002). *The national survey of science and mathematics education: Trends from 1977 to 2000.* Chapel Hill, NC: Horizon Research. Available at: http://www.horizon-research.com/reports/2002/2000survey/trends.php [accessed May 2005].

Sommerville, J., and Yi, Y. (2002). *Aligning K-12 and postsecondary expectations: State policy in transition.* Washington, DC: National Association of System Heads.

Spillane, J.P. (2004). *Standards deviation: How schools misunderstand education policy.* Cambridge, MA: Harvard University Press.

Spillane, J.P., and Zeuli, J.S. (1999). Reform and teaching: Exploring patterns of practice in the context of national and state mathematics reforms. *Educational Evaluation and Policy Analysis, 21*(1), 1-27.

Stecher, B., Barron, S., Chun, T., and Ross, K. (2000). *The effects of the Washington state education reform on schools and classrooms.* (Technical Report No. 525.) Los Angeles: University of California, National Center for Research on Evaluation, Standards, and Student Testing.

The Education Trust. (1999). Ticket to nowhere: The gap between leaving high school and entering college and high-performance jobs. *Thinking K-16, 3*(2), 3-9.

Turner, H. (2004). *CLASS: Creating laboratory access for science students.* Dayton, OH: Wright State University. Available at: http://www.biology.wright.edu/class/main.shtml [accessed Dec. 2004].

Tushnet, N.C., Millsap, M.A., Abdullah-Welsh, N., Brigham, N., Cooley, E., Elliott, J., Johnston, K., Martinez, A., Nierenberg, M., and Rosenblum, S. (2000). *Final report on the evaluation of the National Science Foundation's Instructional Materials Development Program.* Arlington, VA: National Science Foundation.

U.S. Department of Education. (2001). *Digest of education statistics: Tables and figures. Table 153: State requirements for high school graduation, in Carnegie units: 2001.* Washington, DC: Author. Available at: http://www.nces.ed.gov/programs/digest/d01/dt153.asp [accessed April 2005].

U.S. Department of Education. (2003). *Digest of educational statistics 2003.* Washington, D.C: Author. Available at: http://www.nces.ed.gov/programs/digest/d03/tables/dt052.asp [accessed Dec. 2004].

Valverde, G.A., and Schmidt, W.H. (1997, Winter). Refocusing U.S. math and science education. *Issues in Science and Technology Online.* Available at: http://www.issues.org/issues/14.2/schmid.htm [accessed Oct. 2005].

Warren, B., Ballenger, C., Ogonowski, M., Rosebury, A.S., and Hudicourt-Barnes, J. (2001). Rethinking diversity in learning science: The logic of everyday sense-making. *Journal of Research in Science Teaching, 38*(5), 529-552.

Webb, N.L., Kane, J., Kaufman, D., and Yang, J.H. (2001). *Study of the impact of the statewide Systemic Initiatives Program.* Technical report to the National Science Foundation. Madison: Wisconsin Center for Education Research. Available at: http://www.wcer.wisc.edu/SSI/1%20web%20pages2/TechReport2001.htm [accessed May 2005].

Weiss, I.R., Pasley, J.D., Smith, P.S., Banilower, E.R., and Heck, D.J. (2003). *Looking inside the classroom: A study of K-12 mathematics and science education in the United States.* Chapel Hill, NC: Horizon Research.

WestEd. (2004). *PASS science assessment: Partnership for the assessment of standards-based science.* Available at: http://www.wested.org/cs/we/view/serv/9 [accessed Sept. 2004].

3

Laboratory Experiences and Student Learning

Key Points

- *The science learning goals of laboratory experiences include enhancing mastery of science subject matter, developing scientific reasoning abilities, increasing understanding of the complexity and ambiguity of empirical work, developing practical skills, increasing understanding of the nature of science, cultivating interest in science and science learning, and improving teamwork abilities.*
- *The research suggests that laboratory experiences will be more likely to achieve these goals if they (1) are designed with clear learning outcomes in mind, (2) are thoughtfully sequenced into the flow of classroom science instruction, (3) integrate learning of science content and process, and (4) incorporate ongoing student reflection and discussion.*
- *Computer-based representations and simulations of natural phenomena and large scientific databases are more likely to be effective if they are integrated into a thoughtful sequence of classroom science instruction that also includes laboratory experiences.*

In this chapter, the committee first identifies and clarifies the learning goals of laboratory experiences and then discusses research evidence on attainment of those goals. The review of research evidence draws on three major strands of research: (1) cognitive research illuminating how students learn; (2) studies that examine laboratory experiences that stand alone, separate from the flow of classroom science instruction; and (3) research projects that sequence laboratory experiences with other forms of science instruction.[1] We propose the phrase "integrated instructional units" to describe these research and design projects that integrate laboratory experiences within a sequence of science instruction. In the following section of this chapter, we present design principles for laboratory experiences derived from our analysis of these multiple strands of research and suggest that laboratory experiences designed according to these principles are most likely to accomplish their learning goals. Next we consider the role of technology in supporting student learning from laboratory experiences. The chapter concludes with a summary.

GOALS FOR LABORATORY EXPERIENCES

Laboratories have been purported to promote a number of goals for students, most of which are also the goals of science education in general (Lunetta, 1998; Hofstein and Lunetta, 1982). The committee commissioned a paper to examine the definition and goals of laboratory experiences (Millar, 2004) and also considered research reviews on laboratory education that have identified and discussed learning goals (Anderson, 1976; Hofstein and Lunetta, 1982; Lazarowitz and Tamir, 1994; Shulman and Tamir, 1973). While these inventories of goals vary somewhat, a core set remains fairly consistent. Building on these commonly stated goals, the committee developed a comprehensive list of goals for or desired outcomes of laboratory experiences:

- *Enhancing mastery of subject matter.* Laboratory experiences may enhance student understanding of specific scientific facts and concepts and of the way in which these facts and concepts are organized in the scientific disciplines.
- *Developing scientific reasoning.* Laboratory experiences may promote a student's ability to identify questions and concepts that guide scientific

---

[1]There is a larger body of research on how students learn science that is not considered in depth here because the committee's focus is science learning through laboratory experiences. The larger body of research is discussed in the National Research Council (2005) report, *How Students Learn: Science in the Classroom;* it is also considered in an ongoing National Research Council study of science learning in grades K-8.

investigations; to design and conduct scientific investigations; to develop and revise scientific explanations and models; to recognize and analyze alternative explanations and models; and to make and defend a scientific argument. Making a scientific argument includes such abilities as writing, reviewing information, using scientific language appropriately, constructing a reasoned argument, and responding to critical comments.

- *Understanding the complexity and ambiguity of empirical work.* Interacting with the unconstrained environment of the material world in laboratory experiences may help students concretely understand the inherent complexity and ambiguity of natural phenomena. Laboratory experiences may help students learn to address the challenges inherent in directly observing and manipulating the material world, including troubleshooting equipment used to make observations, understanding measurement error, and interpreting and aggregating the resulting data.
- *Developing practical skills.* In laboratory experiences, students may learn to use the tools and conventions of science. For example, they may develop skills in using scientific equipment correctly and safely, making observations, taking measurements, and carrying out well-defined scientific procedures.
- *Understanding of the nature of science.* Laboratory experiences may help students to understand the values and assumptions inherent in the development and interpretation of scientific knowledge, such as the idea that science is a human endeavor that seeks to understand the material world and that scientific theories, models, and explanations change over time on the basis of new evidence.
- *Cultivating interest in science and interest in learning science.* As a result of laboratory experiences that make science "come alive," students may become interested in learning more about science and see it as relevant to everyday life.
- *Developing teamwork abilities.* Laboratory experiences may also promote a student's ability to collaborate effectively with others in carrying out complex tasks, to share the work of the task, to assume different roles at different times, and to contribute and respond to ideas.

Although most of these goals were derived from previous research on laboratory experiences and student learning, the committee identified the new goal of "understanding the complexity and ambiguity of empirical work" to reflect the unique nature of laboratory experiences. Students' direct encounters with natural phenomena in laboratory science courses are inherently more ambiguous and messy than the representations of these phenomena in science lectures, textbooks, and mathematical formulas (Millar, 2004). The committee thinks that developing students' ability to recognize this complexity and develop strategies for sorting through it is an essential goal of laboratory experiences. Unlike the other goals, which coincide with the goals of science education more broadly and may be advanced through lectures, reading, or other forms of science instruction, laboratory experiences may be the only way to advance the goal of helping students understand the complexity and ambiguity of empirical work.

RECENT DEVELOPMENTS IN RESEARCH AND DESIGN OF LABORATORY EXPERIENCES

In reviewing evidence on the extent to which students may attain the goals of laboratory experiences listed above, the committee identified a recent shift in the research. Historically, laboratory experiences have been separate from the flow of classroom science instruction and often lacked clear learning goals. Because this approach remains common today, we refer to these isolated interactions with natural phenomena as "typical" laboratory experiences.[2] Reflecting this separation, researchers often engaged students in one or two experiments or other science activities and then conducted assessments to determine whether their understanding of the science concept underlying the activity had increased. Some studies directly compared measures of student learning following laboratory experiences with measures of student learning following lectures, discussions, videotapes, or other methods of science instruction in an effort to determine which modes of instruction were most effective.

Over the past 10 years, some researchers have shifted their focus. Assuming that the study of the natural world requires opportunities to directly encounter that world, investigators are integrating laboratory experiences and other forms of instruction into instructional sequences in order to help students progress toward science learning goals. These studies draw on principles of learning derived from the rapid growth in knowledge from cognitive research to address the question of *how* to design science instruction, including laboratory experiences, in order to support student learning.

Given the complexity of these teaching and learning sequences, the committee struggled with how best to describe them. Initially, the committee used the term "science curriculum units." However, that term failed to convey the importance of integration in this approach to sequencing laboratory experiences with other forms of teaching and learning. The research reviewed by the committee indicated that these curricula not only integrate laboratory experiences in the flow of science instruction, but also integrate

---

[2]In Chapter 4, we argue that most U.S. high school students currently engage in these typical laboratory experiences.

student learning about both the concepts and processes of science. To reflect these aspects of the new approach, the committee settled on the term "integrated instructional units" in this report.

The following sections briefly describe principles of learning derived from recent research in the cognitive sciences and their application in design of integrated instructional units.

Principles of Learning Informing Integrated Instructional Units

Recent research and development of integrated instructional units that incorporate laboratory experiences are based on a large and growing body of cognitive research. This research has led to development of a coherent and multifaceted theory of learning that recognizes that prior knowledge, context, language, and social processes play critical roles in cognitive development and learning (National Research Council, 1999). Taking each of these factors into account, the National Research Council (NRC) report *How People Learn* identifies four critical principles that support effective learning environments (Glaser, 1994; National Research Council, 1999), and a more recent NRC report, *How Students Learn,* considers these principles as they relate specifically to science (National Research Council, 2005). These four principles are summarized below.

Learner-Centered Environments

The emerging integrated instructional units are designed to be learner-centered. This principle is based on research showing that effective instruction begins with what learners bring to the setting, including cultural practices and beliefs, as well as knowledge of academic content. Taking students' preconceptions into account is particularly critical in science instruction. Students come to the classroom with conceptions of natural phenomena that are based on their everyday experiences in the world. Although these conceptions are often reasonable and can provide satisfactory everyday explanations to students, they do not always match scientific explanations and break down in ways that students often fail to notice. Teachers face the challenge of engaging with these intuitive ideas, some of which are more firmly rooted than others, in order to help students move toward a more scientific understanding. In this way, understanding scientific knowledge often requires a change in—not just an addition to—what students notice and understand about the world (National Research Council, 2005).

Knowledge-Centered Environments

The developing integrated instructional units are based on the principle that learning is enhanced when the environment is knowledge-centered. That is, the laboratory experiences and other instruction included in integrated instructional units are designed to help students learn with understanding, rather than simply acquiring sets of disconnected facts and skills (National Research Council, 1999).

In science, the body of knowledge with which students must engage includes accepted scientific ideas about natural phenomena as well as an understanding of what it means to "do science." These two aspects of science are reflected in the goals of laboratory experiences, which include mastery of subject matter (accepted scientific ideas about phenomena) and several goals related to the processes of science (understanding the complexity of empirical work, development of scientific reasoning). Research on student thinking about science shows a progression of ideas about scientific knowledge and how it is justified. At the first stage, students perceive scientific knowledge as right or wrong. Later, students characterize discrepant ideas and evidence as "mere opinion." Eventually, students recognize scientific knowledge as being justified by evidence derived through rigorous research. Several studies have shown that a large proportion of high school students are at the first stage in their views of scientific knowledge (National Research Council, 2005).

Knowledge-centered environments encourage students to reflect on their own learning progress (metacognition). Learning is facilitated when individuals identify, monitor, and regulate their own thinking and learning. To be effective problem solvers and learners, students need to determine what they already know and what else they need to know in any given situation, including when things are not going as expected. For example, students with better developed metacognitive strategies will abandon an unproductive problem-solving strategy very quickly and substitute a more productive one, whereas students with less effective metacognitive skills will continue to use the same strategy long after it has failed to produce results (Gobert and Clement, 1999). The basic metacognitive strategies include: (1) connecting new information to former knowledge, (2) selecting thinking strategies deliberately, and (3) monitoring one's progress during problem solving.

A final aspect of knowledge-centered learning, which may be particularly relevant to integrated instructional units, is that the practices and activities in which people engage while learning shape what they learn. Transfer (the ability to apply learning in varying situations) is made possible to the extent that knowledge and learning are grounded in multiple contexts. Transfer is more difficult when a concept is taught in a limited set of contexts or through a limited set of activities. By encountering the same concept at work in multiple contexts (such as in laboratory experiences and in discussion), students can develop a deeper understanding of the concept and how it can be used as well as the ability to transfer what has been learned in one context to others (Bransford and Schwartz, 2001).

Assessment to Support Learning

Another important principle of learning that has informed development of integrated instructional units is that assessment can be used to support learning. Cognitive research has shown that feedback is fundamental to learning, but feedback opportunities are scarce in most classrooms. This research indicates that formative assessments provide students with opportunities to revise and improve the quality of their thinking while also making their thinking apparent to teachers, who can then plan instruction accordingly. Assessments must reflect the learning goals of the learning environment. If the goal is to enhance understanding and the applicability of knowledge, it is not sufficient to provide assessments that focus primarily on memory for facts and formulas. The Thinkertools science instructional unit discussed in the following section incorporates this principle, including formative self-assessment tools that help students advance toward several of the goals of laboratory experiences.

Community-Centered Environments

Research has shown that learning is enhanced in a community setting, when students and teachers share norms that value knowledge and participation (see Cobb et al., 2001). Such norms increase people's opportunities and motivation to interact, receive feedback, and learn. Learning is enhanced when students have multiple opportunities to articulate their ideas to peers and to hear and discuss others' ideas. A community-centered classroom environment may not be organized in traditional ways. For example, in science classrooms, the teacher is often the sole authority and arbiter of scientific knowledge, placing students in a relatively passive role (Lemke, 1990). Such an organization may promote students' view that scientific knowledge is a collection of facts about the world, authorized by expert scientists and irrelevant to students' own experience. The instructional units discussed below have attempted to restructure the social organization of the classroom and encourage students and the teacher to interact and learn from each other.

Design of Integrated Instructional Units

The learning principles outlined above have begun to inform design of integrated instructional units that include laboratory experiences with other types of science learning activities. These integrated instructional units were developed through research programs that tightly couple research, design, and implementation in an iterative process. The research programs are beginning to document the details of student learning, development, and interaction when students are given systematic support—or scaffolding—in carefully structured social and cognitive activities. Scaffolding helps to guide students' thinking, so that they can gradually take on more autonomy in carrying out various parts of the activities. Emerging research on these integrated instructional units provides guidance about how to design effective learning environments for real-world educational settings (see Linn, Davis, and Bell, 2004a; Cobb et al., 2003; Design-Based Research Collective, 2003).

Integrated instructional units interweave laboratory experiences with other types of science learning activities, including lectures, reading, and discussion. Students are engaged in framing research questions, designing and executing experiments, gathering and analyzing data, and constructing arguments and conclusions as they carry out investigations. Diagnostic, formative assessments are embedded into the instructional sequences and can be used to gauge student's developing understanding and to promote their self-reflection on their thinking.

With respect to laboratory experiences, these instructional units share two key features. The first is that specific laboratory experiences are carefully selected on the basis of research-based ideas of what students are likely to learn from them. For example, any particular laboratory activity is likely to contribute to learning only if it engages students' current thinking about the target phenomena and is likely to make them critically evaluate their ideas in relation to what they see during the activity. The second is that laboratory experiences are explicitly linked to and integrated with other learning activities in the unit. The assumption behind this second feature is that just because students *do* a laboratory activity, they may not necessarily *understand* what they have done. Nascent research on integrated instructional units suggests that both framing a particular laboratory experience ahead of time and following it with activities that help students make sense of the experience are crucial in using a laboratory experience to support science learning. This "integration" approach draws on earlier research showing that intervention and negotiation with an authority, usually a teacher, was essential to help students make meaning out of their laboratory activities (Driver, 1995).

Examples of Integrated Instructional Units

Scaling Up Chemistry That Applies

Chemistry That Applies (CTA) is a 6-8 week integrated instructional unit designed to help students in grades 8-10 understand the law of conservation of matter. Created by researchers at the Michigan Department of Education (Blakeslee et al., 1993), this instructional unit was one of only a few curricula that were highly rated by American Assocation for the Advancement of Science Project 2061 in its study of middle school science curricula (Kesidou and Roseman, 2002). Student groups explore four chemical reactions—burning, rusting, the decomposition of water, and the volcanic reaction of baking soda and vinegar. They cause these reactions to happen, obtain and record data in individual notebooks, analyze the data, and use evidence-based arguments to explain the data.

The instructional unit engages the students in a carefully structured sequence of hands-on laboratory investigations interwoven with other forms of instruction (Lynch, 2004). Student understanding is "pressed" through many experiences with the reactions and by group and individual pressures to make meaning of these reactions. For example, video transcripts indicate that students engaged in "science talk" during teacher demonstrations and during student experiments.

Researchers at George Washington University, in a partnership with Montgomery County public schools in Maryland, are currently conducting a five-year study of the feasibility of scaling up effective integrated instructional units, including CTA (Lynch, Kuipers, Pyke, and Szesze, in press). In 2001-2002, CTA was implemented in five highly diverse middle schools that were matched with five comparison schools using traditional curriculum materials in a quasi-experimental research design. All 8th graders in the five CTA schools, a total of about 1,500 students, participated in the CTA curriculum, while all 8th graders in the matched schools used the science curriculum materials normally available. Students were given pre- and posttests.

In 2002-2003, the study was replicated in the same five pairs of schools. In both years, students who participated in the CTA curriculum scored significantly higher than comparison students on a posttest. Average scores of students who participated in the CTA curriculum showed higher levels of fluency with the concept of conservation of matter (Lynch, 2004). However, because the concept is so difficult, most students in both the treatment and control group still have misconceptions, and few have a flexible, fully scientific understanding of the conservation of matter. All subgroups of students who were engaged in the CTA curriculum—including low-income students (eligible for free and reduced-price meals), black and Hispanic students, English language learners, and students eligible for special educational services—scored significantly higher than students in the control group on the posttest (Lynch and O'Donnell, 2005). The effect sizes were largest among three subgroups considered at risk for low science achievement, including Hispanic students, low-income students, and English language learners.

Based on these encouraging results, CTA was scaled up to include about 6,000 8th graders in 20 schools in 2003-2004 and 12,000 8th graders in 37 schools in 2004-2005 (Lynch and O'Donnell, 2005).

ThinkerTools

The ThinkerTools instructional unit is a sequence of laboratory experiences and other learning activities that, in its initial version, yielded substantial gains in students' understanding of Newton's laws of motion (White, 1993). Building on these positive results, ThinkerTools was expanded to focus not only on mastery of these laws of motion but also on scientific reasoning and understanding of the nature of science (White and Frederiksen, 1998). In the 10-week unit, students were guided to reflect on their own thinking and learning while they carry out a series of investigations. The integrated instructional unit was designed to help them learn about science processes as well as about the subject of force and motion. The instructional unit supports students as they formulate hypotheses, conduct empirical investigations, work with conceptually analogous computer simulations, and refine a conceptual model for the phenomena. Across the series of investigations, the integrated instructional unit introduces increasingly complex concepts. Formative assessments are integrated throughout the instructional sequence in ways that allow students to self-assess and reflect on core aspects of inquiry and epistemological dimensions of learning.

Researchers investigated the impact of Thinker Tools in 12 7th, 8th, and 9th grade classrooms with 3 teachers and 343 students. The researchers evaluated students' developing understanding of scientific investigations using a pre-post inquiry test. In this assessment, students were engaged in a thought experiment that asked them to conceptualize, design, and think through a hypothetical research study. Gains in scores for students in the reflective self-assessment classes and control classrooms were compared. Results were also broken out by students categorized as high and low achieving, based on performance on a standardized test conducted before the intervention. Students in the reflective self-assessment classes exhibited greater gains on a test of investigative skills. This was especially true for low-achieving students. The researchers further analyzed specific components of the associated scientific processes—formulation of hypotheses, designing an experiment, predicting results, drawing conclusions from made-up results, and relating those conclusions back to the original hypotheses. Students in the reflective-self-assessment classes did better on all of these components than those in control classrooms, especially on the more difficult components (drawing conclusions and relating them to the original hypotheses).

Computer as Learning Partner

Beginning in 1980, a large group of technologists, classroom teachers, and education researchers developed the Computer as Learning Partner (CLP) integrated instructional unit. Over 10 years, the team developed and tested eight versions of a 12-week unit on thermodynamics. Each year, a cohort of about 300 8th grade students participated in a sequence of teaching and learning activities focused primarily on a specific learning goal—enhancing students' understanding of the difference between heat and temperature (Linn, 1997). The project engaged students in a sequence of laboratory experiences supported by computers, discussions, and other forms of science instruction. For example, computer images and words prompted students to make predictions about heat and conductivity and perform experiments using temperature-sensitive probes to confirm or refute their predictions. Students were given tasks related to scientific phenomena affecting their daily lives—such as how to keep a drink cold for lunch or selecting appropriate clothing for hiking in the mountains—as a way to motivate their interest and curiosity. Teachers play an important role in carrying out the curriculum, asking students to critique their own and each others' investigations and encouraging them to reflect on their own thinking.

Over 10 years of study and revision, the integrated instructional unit proved increasingly effective in achieving its stated learning goals. Before the sequenced instruction was introduced, only 3 percent of middle school students could adequately explain the difference between heat and temperature. Eight versions later, about half of the students participating in CLP could explain this difference, representing a 400 percent increase in achievement. In addition, nearly 100 percent of students who participated in the final version of the instructional unit demonstrated understanding of conductors (Linn and Songer, 1991). By comparison, only 25 percent of a group of undergraduate chemistry students at the University of California at Berkeley could adequately explain the difference between heat and temperature. A longitudinal study comparing high school seniors who participated in the thermodynamics unit in middle school with seniors who had received more traditional middle school science instruction found a 50 percent improvement in CLP students' performance in distinguishing between heat and temperature (Linn and Hsi, 2000)

Participating in the CLP instructional unit also increased students' interest in science. Longitudinal studies of CLP participants revealed that, among those who went on to take high school physics, over 90 percent thought science was relevant to their lives. And 60 percent could provide examples of scientific phenomena in their daily lives. By comparison, only 60 percent of high school physics students who had not participated in the unit during middle school thought science was relevant to their lives, and only 30 percent could give examples in their daily lives (Linn and Hsi, 2000).

EFFECTIVENESS OF LABORATORY EXPERIENCES

Description of the Literature Review

The committee's review of the literature on the effectiveness of laboratory experiences considered studies of typical laboratory experiences and emerging research focusing on integrated instructional units. In reviewing both bodies of research, we aim to specify how laboratory experiences can further each of the science learning goals outlined at the beginning of this chapter.

Limitations of the Research

Our review was complicated by weaknesses in the earlier research on typical laboratory experiences, isolated from the stream of instruction (Hofstein and Lunetta, 1982). First, the investigators do not agree on a precise definition of the "laboratory" experiences under study. Second, many studies were weak in the selection and control of variables. Investigators failed to examine or report important variables relating to student abilities and attitudes. For example, they failed to note students' prior laboratory experiences. They also did not give enough attention to extraneous factors that might affect student outcomes, such as instruction outside the laboratory. Third, the studies of typical laboratory experiences usually involved a small group of students with little diversity, making it difficult to generalize the results to the large, diverse population of U.S. high schools today. Fourth, investigators did not give enough attention to the adequacy of the instruments used to measure student outcomes. As an example, paper and pencil tests that focus on testing mastery of subject matter, the most frequently used assessment, do not capture student attainment of all of the goals we have identified. Such tests are not able to measure student progress toward goals that may be unique to laboratory experiences, such as developing scientific reasoning, understanding the complexity and ambiguity of empirical work, and development of practical skills.

Finally, most of the available research on typical laboratory experiences does not fully describe these activities. Few studies have examined teacher behavior, the classroom learning environment, or variables identifying teacher-student interaction. In addition, few recent studies have focused on laboratory manuals—both what is in them and how they are used. Research on the intended design of laboratory experiences, their implementation, and whether the implementation resembles the initial design would provide the understanding needed to guide improvements in laboratory instruction. However, only a few studies of typical laboratory experiences have measured the effectiveness of particular laboratory experiences in terms of both the extent to which their activities match those that the teacher intended and the extent to which the students' learning matches the learning objectives of the activity (Tiberghien, Veillard, Le Marchal, Buty, and Millar, 2000).

We also found weaknesses in the evolving research on integrated instructional units. First, these new units tend to be hothouse projects; researchers work intensively with teachers to construct atypical learning environments. While some have been developed and studied over a number of years and iterations, they usually involve relatively small samples of students. Only now are some of these efforts expanding to a scale that will allow robust generalizations about their value and how best to implement them. Second, these integrated instructional units have not been designed specifically to contrast some version of laboratory or practical experience with a lack of such experience. Rather, they assume that educational interventions are complex, systemic "packages" (Salomon, 1996) involving many interactions that may influence specific outcomes, and that science learning requires some opportunities for direct engagement with natural phenomena. Researchers commonly aim to document the complex interactions between and among students, teachers, laboratory materials, and equipment in an effort to develop profiles of successful interventions (Cobb et al., 2003; Collins, Joseph, and Bielaczyc, 2004; Design-Based Research Collective, 2003). These newer studies focus on how to sequence laboratory experiences and other forms of science instruction to support students' science learning.

Scope of the Literature Search

A final note on the review of research: the scope of our study did not allow for an in-depth review of all of the individual studies of laboratory education conducted over the past 30 years. Fortunately, three major reviews of the literature from the 1970s, 1980s, and 1990s are available (Lazarowitz and Tamir, 1994; Lunetta, 1998; Hofstein and Lunetta, 2004). The committee relied on these reviews in our analysis of studies published before 1994. To identify studies published between 1994 and 2004, the committee searched electronic databases.

To supplement the database search, the committee commissioned three experts to review the nascent body of research on integrated instructional units (Bell, 2005; Duschl, 2004; Millar, 2004). We also invited researchers who are currently developing, revising, and studying the effectiveness of integrated instructional units to present their findings at committee meetings (Linn, 2004; Lynch, 2004).

All of these activities yielded few studies that focused on the high school level and were conducted in the United States. For this reason, the committee expanded the range of the literature considered to include some studies targeted at middle school and some international studies. We included studies at the elementary through postsecondary levels as well as studies of teachers' learning in our analysis. In drawing conclusions from studies that were not conducted at the high school level, the committee took into consideration the extent to which laboratory experiences in high school differ from those in elementary and postsecondary education. Developmental differences among students, the organizational structure of schools, and the preparation of teachers are a few of the many factors that vary by school level and that the committee considered in making inferences from the available research. Similarly, when deliberating on studies conducted outside the United States, we considered differences in the science curriculum, the organization of schools, and other factors that might influence the outcomes of laboratory education.

Mastery of Subject Matter

Evidence from Research on Typical Laboratory Experiences

Claims that typical laboratory experiences help students master science content rest largely on the argument that opportunities to directly interact with, observe, and manipulate materials will help students to better grasp difficult scientific concepts. It is believed that these experiences will force students to confront their misunderstandings about phenomena and shift toward more scientific understanding.

Despite these claims, there is almost no direct evidence that typical laboratory experiences that are isolated from the flow of science instruction are particularly valuable for learning specific scientific content (Hofstein and Lunetta, 1982, 2004; Lazarowitz and Tamir, 1994). White (1996) points out that many major reviews of science education from the 1960s and 1970s indicate that laboratory work does little to improve understanding of science content as measured by paper and pencil tests, and later studies from the 1980s and early 1990s do not challenge this view. Other studies indicate that typical laboratory experiences are no more effective in helping students master science subject matter than demonstrations in high school biology (Coulter, 1966), demonstration and discussion (Yager, Engen, and Snider, 1969), and viewing filmed experiments in chemistry (Ben-Zvi, Hofstein, Kempa, and Samuel, 1976). In contrast to most of the research, a single comparative study (Freedman, 2002) found that students who received regular laboratory instruction over the course of a school year performed better on a test of physical science knowledge than a control group of students who took a similar physical science course without laboratory activities.

Clearly, most of the evidence does not support the argument that typical laboratory experiences lead to improved learning of science content. More specifically, concrete experiences with phenomena alone do not appear to force students to confront their misunderstandings and reevaluate their own assumptions. For example, VandenBerg, Katu, and Lunetta (1994) reported, on the basis of clinical studies with individual students, that hands-on activities with introductory electricity materials facilitated students' understanding of the relationships among circuit elements and variables. The carefully selected practical activities created conceptual conflict in students' minds—a first step toward changing their naïve ideas about electricity. However, the students remained unable to develop a fully scientific mental model of a circuit system. The authors suggested that greater engagement with conceptual organizers, such as analogies and concept maps, could have helped students develop more scientific understandings of basic electricity. Several researchers, including Dupin and Joshua (1987), have reported similar findings. Studies indicate that students often hold beliefs so intensely that even their observations in the laboratory are strongly influenced by those beliefs (Champagne, Gunstone, and Klopfer, 1985, cited in Lunetta, 1998; Linn, 1997). Students tend to adjust their observations to fit their current beliefs rather than change their beliefs in the face of conflicting observations.

Evidence from Research on Integrated Instructional Units

Current integrated instructional units build on earlier studies that found integration of laboratory experiences with other instructional activities enhanced mastery of subject matter (Dupin and Joshua, 1987; White and Gunstone, 1992, cited in Lunetta, 1998). A recent review of these and other studies concluded (Hofstein and Lunetta, 2004, p. 33):

> When laboratory experiences are integrated with other metacognitive learning experiences such as "predict-observe-explain" demonstrations (White and Gunstone, 1992) and when they incorporate the manipulation of ideas instead of simply materials and procedures, they can promote the learning of science.

Integrated instructional units often focus on complex science topics that are difficult for students to understand. Their design is based on research on students' intuitive conceptions of a science topic and how those conceptions differ from scientific conceptions. Students' ideas often do not match the scientific understanding of a phenomenon and, as noted previously, these intuitive notions are resistant to change. For this reason, the sequenced units incorporate instructional activities specifically designed to confront intuitive conceptions and provide an environment in which students can construct normative conceptions. The role of laboratory experiences is to emphasize the discrepancies between students' intuitive ideas about the topic and scientific ideas, as well as to support their construction of normative understanding. In order to help students link formal, scientific concepts to real phenomena, these units include a sequence of experiences that will push them to question their intuitive and often inaccurate ideas.

Emerging studies indicate that exposure to these integrated instructional units leads to demonstrable gains in student mastery of a number of science topics in comparison to more traditional approaches. In physics, these subjects include Newtonian mechanics (Wells, Hestenes, and Swackhamer, 1995; White, 1993); thermodynamics (Songer and Linn, 1991); electricity (Shaffer and McDermott, 1992); optics (Bell and Linn, 2000; Reiner, Pea, and Shulman, 1995); and matter (Lehrer, Schauble, Strom, and Pligge, 2001; Smith, Maclin, Grosslight, and Davis, 1997; Snir, Smith, and Raz, 2003). Integrated instructional units in biology have enhanced student mastery of genetics (Hickey, Kindfield, Horwitz, and Christie, 2003) and natural selection (Reiser et al., 2001). A chemistry unit has led to gains in student understanding of stoichiometry (Lynch, 2004). Many, but not all, of these instructional units combine computer-based simulations of the phenomena under study with direct interactions with these phenomena. The role of technology in providing laboratory experiences is described later in this chapter.

Developing Scientific Reasoning

While philosophers of science now agree that there is no single scientific method, they do agree that a number of reasoning skills are critical to research across the natural sciences. These reasoning skills include identifying questions and concepts that guide scientific investigations, designing and conducting scientific investigations, developing and revising scientific explanations and models, recognizing and analyzing alternative explanations and models, and making and defending a scientific argument. It is not necessarily the case that these skills are sequenced in a particular way or used in every scientific investigation. Instead, they are representative of the abilities that both scientists and students need to investigate the material world and make meaning out of those investigations. Research on children's and adults' scientific reasoning (see the review by Zimmerman, 2000) suggests that effective experimentation is difficult for most people and not learned without instructional support.

Evidence from Research on Typical Laboratory Experiences

Early research on the development of investigative skills suggested that students could learn aspects of scientific reasoning through typical laboratory instruction in college-level physics (Reif and St. John, 1979, cited in Hofstein and Lunetta, 1982) and in high school and college biology (Raghubir, 1979; Wheatley, 1975, cited in Hofstein and Lunetta, 1982).

More recent research, however, suggests that high school and college science teachers often emphasize laboratory procedures, leaving little time for discussion of how to plan an investigation or interpret its results (Tobin, 1987; see Chapter 4). Taken as a whole, the evidence indicates that typical laboratory work promotes only a few aspects of the full process of scientific reasoning—making observations and organizing, communicating, and interpreting data gathered from these observations. Typical laboratory experiences appear to have little effect on more complex aspects of scientific reasoning, such as the capacity to formulate research questions, design experiments, draw conclusions from observational data, and make inferences (Klopfer, 1990, cited in White, 1996).

Evidence from Research on Integrated Instructional Units

Research developing from studies of integrated instructional units indicates that laboratory experiences can play an important role in developing all aspects of scientific reasoning, including the more complex aspects, if the laboratory experiences are integrated with small group discussion, lectures, and other forms of science instruction. With carefully designed instruction that incorporates opportunities to conduct investigations and reflect on the results, students as young as 4th and 5th grade can develop sophisticated scientific thinking (Lehrer and Schauble, 2004; Metz, 2004). Kuhn and colleagues have shown that 5th graders can learn to experiment effectively, albeit in carefully controlled domains and with extended supervised practice (Kuhn, Schauble, and Garcia-Mila, 1992). Explicit instruction on the purposes of experiments appears necessary to help 6th grade students design them well (Schauble, Giaser, Duschl, Schulze, and John, 1995). These studies suggest that laboratory experiences must be carefully designed to support the development of scientific reasoning.

Given the difficulty most students have with reasoning scientifically, a number of instructional units have focused on this goal. Evidence from several studies indicates that, with the appropriate scaffolding provided in these units, students can successfully reason scientifically. They can learn to design experiments (Schauble et al., 1995; White and Frederiksen, 1998), make predictions (Friedler, Nachmias, and Linn, 1990), and interpret and explain data (Bell and Linn, 2000; Coleman, 1998; Hatano and Inagaki, 1991; Meyer and Woodruff, 1997; Millar, 1998; Rosebery, Warren, and Conant, 1992; Sandoval and Millwood, 2005). Engagement with these instructional units has been shown to improve students' abilities to recognize discrepancies between predicted and observed outcomes (Friedler et al., 1990) and to design good experiments (Dunbar, 1993; Kuhn et al., 1992; Schauble et al., 1995; Schauble, Klopfer, and Raghavan, 1991).

Integrated instructional units seem especially beneficial in developing scientific reasoning skills among lower ability students (White and Frederiksen, 1998).

Recently, research has focused on an important element of scientific reasoning—the ability to construct scientific arguments. Developing, revising, and communicating scientific arguments is now recognized as a core scientific practice (Driver, Newton, and Osborne, 2000; Duschl and Osborne, 2002). Laboratory experiences play a key role in instructional units designed to enhance students' argumentation abilities, because they provide both the impetus and the data for constructing scientific arguments. Such efforts have taken many forms. For example, researchers working with young Haitian-speaking students in Boston used the students' own interests to develop scientific investigations. Students designed an investigation to determine which school drinking fountain had the best-tasting water. The students designed data collection protocols, collected and analyzed their data, and then argued about their findings (Rosebery et al., 1992). The Knowledge Integration Environment project asked middle school students to examine a common set of evidence to debate competing hypotheses about light propagation. Overall, most students learned the scientific concept (that light goes on forever), although those who made better arguments learned more than their peers (Bell and Linn, 2000). These and other examples (e.g., Sandoval and Millwood, 2005) show that students in middle and high school can learn to argue scientifically, by learning to coordinate theoretical claims with evidence taken from their laboratory investigations.

Developing Practical Skills

Evidence from Research on Typical Laboratory Experiences

Science educators and researchers have long claimed that learning practical laboratory skills is one of the important goals for laboratory experiences and that such skills may be attainable only through such experiences (White, 1996; Woolnough, 1983). However, development of practical skills has been measured in research less frequently than mastery of subject matter or scientific reasoning. Such practical outcomes deserve more attention, especially for laboratory experiences that are a critical part of vocational or technical training in some high school programs. When a primary goal of a program or course is to train students for jobs in laboratory settings, they must have the opportunity to learn to use and read sophisticated instruments and carry out standardized experimental procedures. The critical questions about acquiring these skills through laboratory experiences may not be whether laboratory experiences help students learn them, but how the experiences can be constructed so as to be most effective in teaching such skills.

Some research indicates that typical laboratory experiences specifically focused on learning practical skills can help students progress toward other goals. For example, one study found that students were often deficient in the simple skills needed to successfully carry out typical laboratory activities, such as using instruments to make measurements and collect accurate data (Bryce and Robertson, 1985). Other studies indicate that helping students to develop relevant instrumentation skills in controlled "prelab" activities can reduce the probability that important measurements in a laboratory experience will be compromised due to students' lack of expertise with the apparatus (Beasley, 1985; Singer, 1977). This research suggests that development of practical skills may increase the probability that students will achieve the intended results in laboratory experiences. Achieving the intended results of a laboratory activity is a necessary, though not sufficient, step toward effectiveness in helping students attain laboratory learning goals.

Some research on typical laboratory experiences indicates that girls handle laboratory equipment less frequently than boys, and that this tendency is associated with less interest in science and less self-confidence in science ability among girls (Jovanovic and King, 1998). It is possible that helping girls to develop instrumentation skills may help them to participate more actively and enhance their interest in learning science.

Evidence from Research on Integrated Instructional Units

Studies of integrated instructional units have not examined the extent to which engagement with these units may enhance practical skills in using laboratory materials and equipment. This reflects an instructional emphasis on helping students to learn scientific ideas with real understanding and on developing their skills at investigating scientific phenomena, rather than on particular laboratory techniques, such as taking accurate measurements or manipulating equipment. There is no evidence to suggest that students do not learn practical skills through integrated instructional units, but to date researchers have not assessed such practical skills.

Understanding the Nature of Science

Throughout the past 50 years, studies of students' epistemological beliefs about science consistently show that most of them have naïve views about the nature of scientific knowledge and how such knowledge is constructed and evaluated by scientists over time (Driver, Leach, Millar, and Scott, 1996; Lederman, 1992). The general public understanding of science is similarly inaccurate. Firsthand experience with science is often seen as a key way to advance students' understanding of and appreciation for the conventions of science. Laboratory experiences are considered the primary mechanism for providing firsthand experience and are therefore assumed to improve students' understanding of the nature of science.

Evidence from Research on Typical Laboratory Experiences

Research on student understanding of the nature of science provides little evidence of improvement with science instruction (Lederman, 1992; Driver et al., 1996). Although much of this research historically did not examine details of students' laboratory experiences, it often included very large samples of science students and thus arguably captured typical laboratory experiences (research from the late 1950s through the 1980s is reviewed by Lederman, 1992). There appear to be developmental trends in students' understanding of the relations between experimentation and theory-building. Younger students tend to believe that experiments yield direct answers to questions; during middle and high school, students shift to a vague notion of experiments being tests of ideas. Only a small number of students appear to leave high school with a notion of science as model-building and experimentation, in an ongoing process of testing and revision (Driver et al., 1996; Carey and Smith, 1993; Smith et al., 2000). The conclusion that most experts draw from these results is that the isolated nature and rote procedural focus of typical laboratory experiences inhibits students from developing robust conceptions of the nature of science. Consequently, some have argued that the nature of science must be an explicit target of instruction (Khishfe and Abd-El-Khalick, 2002; Lederman, Abd-El-Khalick, Bell, and Schwartz, 2002).

Evidence from Research on Integrated Instructional Units

As discussed above, there is reasonable evidence that integrated instructional units help students to learn processes of scientific inquiry. However, such instructional units do not appear, on their own, to help students develop robust conceptions of the nature of science. One large-scale study of a widely available inquiry-oriented curriculum, in which integrated instructional units were an explicit feature, showed no significant change in students' ideas about the nature of science after a year's instruction (Meichtry, 1993). Students engaged in the BGuILE science instructional unit showed no gains in understanding the nature of science from their participation, and they seemed not even to see their experience in the unit as necessarily related to professional science (Sandoval and Morrison, 2003). These findings and others have led to the suggestion that the nature of science must be an explicit target of instruction (Lederman et al., 2002).

There is evidence from the ThinkerTools science instructional unit that by engaging in reflective self-assessment on their own scientific investigations, students gained a more sophisticated understanding of the nature of science than matched control classes who used the curriculum without the ongoing monitoring and evaluation of their own and others' research (White and Frederiksen, 1998). Students who engaged in the reflective assessment process "acquire knowledge of the forms that scientific laws, models, and theories can take, and of how the development of scientific theories is related to empirical evidence" (White and Frederiksen, 1998, p. 92). Students who participated in the laboratory experiences and other learning activities in this unit using the reflective assessment process were less likely to "view scientific theories as immutable and never subject to revision" (White and Frederiksen, 1998, p. 72). Instead, they saw science as meaningful and explicable. The ThinkerTools findings support the idea that attention to nature of science issues should be an explicit part of integrated instructional units, although even with such attention it remains difficult to change students' ideas (Khishfe and Abd-el-Khalick, 2002).

A survey of several integrated instructional units found that they seem to bridge the "language gap" between science in school and scientific practice (Duschl, 2004). The units give students "extended opportunities to explore the relationship between evidence and explanation," helping them not only to develop new knowledge (mastery of subject matter), but also to evaluate claims of scientific knowledge, reflecting a deeper understanding of the nature of science (Duschl, 2004). The available research leaves open the question of whether or not these experiences help students to develop an explicit, reflective conceptual framework about the nature of science.

Cultivating Interest in Science and Interest in Learning Science

Evidence from Research on Typical Laboratory Experiences

Studies of the effect of typical laboratory experiences on student interest are much rarer than those focusing on student achievement or other cognitive outcomes (Hofstein and Lunetta, 2004; White, 1996). The number of studies that address interest, attitudes, and other affective outcomes has decreased over the past decade, as researchers have focused almost exclusively on cognitive outcomes (Hofstein and Lunetta, 2004). Among the few studies available, the evidence is mixed. Some studies indicate that laboratory experiences lead to more positive attitudes (Renner, Abraham, and Birnie, 1985; Denny and Chennell, 1986). Other studies show no relation between laboratory experiences and affect (Ato and Wilkinson, 1986; Freedman, 2002), and still others report laboratory experiences turned students away from science (Holden, 1990; Shepardson and Pizzini, 1993).

There are, however, two apparent weaknesses in studies of interest and attitude (Hofstein and Lunetta, 1982). One is that researchers often do not carefully define interest and how it should be measured. Consequently, it is unclear if students simply reported liking laboratory activities more than other classroom activities, or if laboratory activities engendered more interest in science as a field, or in taking science courses, or something else. Similarly, studies may report increased positive attitudes toward science from students' participation in laboratory experiences, without clear description of what attitudes were measured, how large the changes were, or whether changes persisted over time.

Student Perceptions of Typical Laboratory Experiences

Students' perceptions of laboratory experiences may affect their interest and engagement in science, and some studies have examined those perceptions. Researchers have found that students often do not have clear ideas about the general or specific purposes of their work in typical science laboratory activities (Chang and Lederman, 1994) and that their understanding of the goals of lessons frequently do not match their teachers' goals for the same lessons (Hodson, 1993; Osborne and Freyberg, 1985; Wilkenson and Ward, 1997). When students do not understand the goals of experiments or laboratory investigations, negative consequences for learning occur (Schauble et al., 1995). In fact, students often do not make important connections between the purpose of a typical laboratory investigation and the design of the experiments. They do not connect the experiment with what they have done earlier, and they do not note the discrepancies among their own concepts, the concepts of their peers, and those of the science community (Champagne et al., 1985; Eylon and Linn, 1988; Tasker, 1981). As White (1998) notes, "to many students, a 'lab' means manipulating equipment but not manipulating ideas." Thus, in considering how laboratory experiences may contribute to students' interest in science and to other learning goals, their perceptions of those experiences must be considered.

A series of studies using the Science Laboratory Environment Inventory (SLEI) has demonstrated links between students' perceptions of laboratory experiences and student outcomes (Fraser, McRobbie, and Giddings, 1993; Fraser, Giddings, and McRobbie, 1995; Henderson, Fisher, and Fraser, 2000; Wong and Fraser, 1995). The SLEI, which has been validated cross-nationally, measures five dimensions of the laboratory environment: student cohesiveness, open-endedness, integration, rule clarity, and material environment (see Table 3-1 for a description of each scale). Using the SLEI, researchers have studied students' perceptions of chemistry and biology laboratories in several countries, including the United States. All five dimensions appear to be positively related with student attitudes, although the

TABLE 3-1 Descriptive Information for the Science Laboratory Environment Inventory

Scale Name	Description
Student cohesiveness	Extent to which students know, help, and are supportive of one another
Open-endedness	Extent to which the laboratory activities emphasize an open-ended, divergent approach to experimentation
Integration	Extent to which laboratory activities are integrated with nonlaboratory and theory classes
Rule clarity	Extent to which behavior in the laboratory is guided by formal rules
Material environment	Extent to which the laboratory equipment and materials are adequate

SOURCE: Henderson, Fisher, and Fraser (2000). Reprinted with permission of Wiley-Liss, Inc., a subsidiary of John Wiley & Sons, Inc.

relation of open-endedness with attitudes seems to vary with student population. In some populations, there is a negative relation to attitudes (Fraser et al., 1995) and to some cognitive outcomes (Henderson et al., 2000).

Research using the SLEI indicates that positive student attitudes are particularly strongly associated with cohesiveness (the extent to which students know, help, and are supportive of one another) and integration (the extent to which laboratory activities are integrated with nonlaboratory and theory classes) (Fraser et al.,1995; Wong and Fraser, 1995). Integration also shows a positive relation to students' cognitive outcomes (Henderson et al., 2000; McRobbie and Fraser, 1993).

Evidence from Research on Integrated Instructional Units

Students' interest and attitudes have been measured less often than other goals of laboratory experiences in studies of integrated instructional units. When evidence is available, it suggests that students who participate in these units show greater interest in and more positive attitudes toward science. For example, in a study of ThinkerTools, completion of projects was used as a measure of student interest. The rate of submitting completed projects was higher for students in the ThinkerTools curriculum than for those in traditional instruction. This was true for all grades and ability levels (White and

Frederiksen, 1998). This study also found that students' ongoing evaluation of their own and other students' thinking increased motivation and self-confidence in their individual ability: students who participated in this ongoing evaluation not only turned in their final project reports more frequently, but they were also less likely to turn in reports that were identical to their research partner's.

Participation in the ThinkerTools instructional unit appears to change students' attitudes toward learning science. After completing the integrated instructional unit, fewer students indicated that "being good at science" was a result of inherited traits, and fewer agreed with the statement, "In general, boys tend to be naturally better at science than girls." In addition, more students indicated that they preferred taking an active role in learning science, rather than simply being told the correct answer by the teacher (White and Frederiksen, 1998).

Researchers measured students' engagement and motivation to master the complex topic of conservation of matter as part of the study of CTA. Students who participated in the CTA curriculum had higher levels of basic engagement (active participation in activities) and were more likely to focus on learning from the activities than students in the control group (Lynch et al., in press). This positive effect on engagement was especially strong among low-income students. The researchers speculate, "perhaps as a result of these changes in engagement and motivation, they learned more than if they had received the standard curriculum" (Lynch et al., in press).

Students who participated in CLP during middle school, when surveyed years later as high school seniors, were more likely to report that science is relevant to their lives than students who did not participate (Linn and Hsi, 2000). Further research is needed to illuminate which aspects of this instructional unit contribute to increased interest.

Developing Teamwork Abilities

Evidence from Research on Typical Laboratory Experiences

Teamwork and collaboration appear in research on typical laboratory experiences in two ways. First, working in groups is seen as a way to enhance student learning, usually with reference to literature on cooperative learning or to the importance of providing opportunities for students to discuss their ideas. Second and more recently, attention has focused on the ability to work in groups as an outcome itself, with laboratory experiences seen as an ideal opportunity to develop these skills. The focus on teamwork as an outcome is usually linked to arguments that this is an essential skill for workers in the 21st century (Partnership for 21st Century Skills, 2003).

Evidence from Research on Integrated Instructional Units

There is considerable evidence that collaborative work can help students learn, especially if students with high ability work with students with low ability (Webb and Palincsar, 1996). Collaboration seems especially helpful to lower ability students, but only when they work with more knowledgeable peers (Webb, Nemer, Chizhik, and Sugrue, 1998). Building on this research, integrated instructional units engage students in small-group collaboration as a way to encourage them to connect what they know (either from their own experiences or from prior instruction) to their laboratory experiences. Often, individual students disagree about prospective answers to the questions under investigation or the best way to approach them, and collaboration encourages students to articulate and explain their reasoning. A number of studies suggest that such collaborative investigation is effective in helping students to learn targeted scientific concepts (Coleman, 1998; Roschelle, 1992).

Extant research lacks specific assessment of the kinds of collaborative skills that might be learned by individual students through laboratory work. The assumption appears to be that if students collaborate and such collaborations are effective in supporting their conceptual learning, then they are probably learning collaborative skills, too.

Overall Effectiveness of Laboratory Experiences

The two bodies of research—the earlier research on typical laboratory experiences and the emerging research on integrated instructional units—yield different findings about the effectiveness of laboratory experiences in advancing the goals identified by the committee. In general, the nascent body of research on integrated instructional units offers the promise that laboratory experiences embedded in a larger stream of science instruction can be more effective in advancing these goals than are typical laboratory experiences (see Table 3-2).

Research on the effectiveness of typical laboratory experiences is methodologically weak and fragmented. The limited evidence available suggests that typical laboratory experiences, by themselves, are neither better nor worse than other methods of science instruction for helping students master science subject matter. However, more recent research indicates that integrated instructional units enhance students' mastery of subject matter. Studies have demonstrated increases in student mastery of complex topics in physics, chemistry, and biology.

Typical laboratory experiences appear, based on the limited research available, to support some aspects of scientific reasoning; however, typical laboratory experiences alone are not sufficient for promoting more sophisticated scientific reasoning abilities, such as asking appropriate questions,

TABLE 3-2 Attainment of Educational Goals in Typical Laboratory Experiences and Integrated Instructional Units

Goal	Typical Laboratory Experiences	Integrated Instructional Units
Mastery of subject matter	No better or worse than other modes of instruction	Increased mastery compared with other modes of instruction
Scientific reasoning	Aids development of some aspects	Aids development of more sophisticated aspects
Understanding of the nature of science	Little improvement	Some improvement when explicitly targeted at this goal
Interest in science	Some evidence of increased interest	Greater evidence of increased interest
Understanding the complexity and ambiguity of empirical work	Inadequate evidence	Inadequate evidence
Development of practical skills	Inadequate evidence	Inadequate evidence
Development of teamwork skills	Inadequate evidence	Inadequate evidence

designing experiments, and drawing inferences. Research on integrated instructional units provides evidence that the laboratory experiences and other forms of instruction they include promote development of several aspects of scientific reasoning, including the ability to ask appropriate questions, design experiments, and draw inferences.

The evidence indicates that typical laboratory experiences do little to increase students' understanding of the nature of science. In contrast, some studies find that participating in integrated instructional units that are designed specifically with this goal in mind enhances understanding of the nature of science.

The available research suggests that typical laboratory experiences can play a role in enhancing students' interest in science and in learning science. There is evidence that engagement with the laboratory experiences and other learning activities included in integrated instructional units enhances students' interest in science and motivation to learn science.

In sum, the evolving research on integrated instructional units provides evidence of increases in students' understanding of subject matter, development of scientific reasoning, and interest in science, compared with students who received more traditional forms of science instruction. Studies conducted to date also suggest that the units are effective in helping diverse groups of students attain these three learning goals. In contrast, the earlier research on typical laboratory experiences indicates that such typical laboratory experiences are neither better nor worse than other forms of science instruction in supporting student mastery of subject matter. Typical laboratory experiences appear to aid in development of only some aspects of scientific reasoning, and they appear to play a role in enhancing students' interest in science and in learning science.

Due to a lack of available studies, the committee was unable to draw conclusions about the extent to which either typical laboratory experiences or laboratory experiences incorporated into integrated instructional units might advance the other goals identified at the beginning of this chapter—enhancing understanding of the complexity and ambiguity of empirical work, acquiring practical skills, and developing teamwork skills.

PRINCIPLES FOR DESIGN OF EFFECTIVE LABORATORY EXPERIENCES

The three bodies of research we have discussed—research on how people learn, research on typical laboratory experiences, and developing research on how students learn in integrated instructional units—yield information that promises to inform the design of more effective laboratory experiences.

The committee considers the emerging evidence sufficient to suggest four general principles that can help laboratory experiences achieve the goals outlined above. It must be stressed, however, that research to date has not described in much detail how these principles can be implemented nor how each principle might relate to each of the educational goals of laboratory experiences.

Clearly Communicated Purposes

Effective laboratory experiences have clear learning goals that guide the design of the experience. Ideally these goals are clearly communicated to students. Without a clear understanding of the purposes of a laboratory activity, students seem not to get much from it. Conversely, when the purposes of a laboratory activity are clearly communicated by teachers to students, then students seem capable of understanding them and carrying them out. There seems to be no compelling evidence that particular purposes are more understandable to students than others.

Sequenced into the Flow of Instruction

Effective laboratory experiences are thoughtfully sequenced into the flow of classroom science instruction. That is, they are explicitly linked to what has come before and what will come after. A common theme in reviews of laboratory practice in the United States is that laboratory experiences are presented to students as isolated events, unconnected with other aspects of classroom work. In contrast, integrated instructional units embed laboratory experiences with other activities that build on the laboratory experiences and push students to reflect on and better understand these experiences. The way a particular laboratory experience is integrated into a flow of activities should be guided by the goals of the overall sequence of instruction and of the particular laboratory experience.

Integrated Learning of Science Concepts and Processes

Research in the learning sciences (National Research Council, 1999, 2001) strongly implies that conceptual understanding, scientific reasoning, and practical skills are three capabilities that are not mutually exclusive. An educational program that partitions the teaching and learning of content from the teaching and learning of process is likely to be ineffective in helping students develop scientific reasoning skills and an understanding of science as a way of knowing. The research on integrated instructional units, all of which intertwine exploration of content with process through laboratory experiences, suggests that integration of content and process promotes attainment of several goals identified by the committee.

Ongoing Discussion and Reflection

Laboratory experiences are more likely to be effective when they focus students more on discussing the activities they have done during their laboratory experiences and reflecting on the meaning they can make from them, than on the laboratory activities themselves. Crucially, the focus of laboratory experiences and the surrounding instructional activities should not simply be on confirming presented ideas, but on developing explanations to make sense of patterns of data. Teaching strategies that encourage students to articulate their hypotheses about phenomena prior to experimentation and to then reflect on their ideas after experimentation are demonstrably more successful at supporting student attainment of the goals of mastery of subject matter, developing scientific reasoning, and increasing interest in science and science learning. At the same time, opportunities for ongoing discussion and reflection could potentially support students in developing teamwork skills.

COMPUTER TECHNOLOGIES AND LABORATORY EXPERIENCES

From scales to microscopes, technology in many forms plays an integral role in most high school laboratory experiences. Over the past two decades, personal computers have enabled the development of software specifically designed to help students learn science, and the Internet is an increasingly used tool for science learning and for science itself. This section examines the role that computer technologies now and may someday play in science learning in relation to laboratory experiences. Certain uses of computer technology can be seen as laboratory experiences themselves, according to the committee's definition, to the extent that they allow students to interact with data drawn directly from the world. Other uses, less clearly laboratory experiences in themselves, provide certain features that aid science learning.

Computer Technologies Designed to Support Learning

Researchers and science educators have developed a number of software programs to support science learning in various ways. In this section, we summarize what we see as the main ways in which computer software can support science learning through providing or augmenting laboratory experiences.

Scaffolded Representations of Natural Phenomena

Perhaps the most common form of science education software are programs that enable students to interact with carefully crafted models of natural phenomena that are difficult to see and understand in the real world and have proven historically difficult for students to understand. Such programs are able to show conceptual interrelationships and connections between theoretical constructs and natural phenomena through the use of multiple, linked representations. For example, velocity can be linked to acceleration and position in ways that make the interrelationships understandable to students (Roschelle, Kaput, and Stroup, 2000). Chromosome genetics can be linked to changes in pedigrees and populations (Horowitz, 1996). Molecular chemical representations can be linked to chemical equations (Kozma, 2003).

In the ThinkerTools integrated instructional unit, abstracted representations of force and motion are provided for students to help them "see" such ideas as force, acceleration, and velocity in two dimensions (White, 1993; White and Frederiksen, 1998). Objects in the ThinkerTools microworld are represented as simple, uniformly sized "dots" to avoid students becoming confused about the idea of center of mass. Students use the microworld to solve various problems of motion in one or two dimensions, using the computer keyboard to apply forces to dots to move them along specified paths. Part of the key to the software's guidance is that it provides representations of forces and accelerations in which students can see change in response to their actions. A "dot trace," for example, shows students how applying more force affects an object's acceleration in a predictable way. A "vector cross" represents the individual components of forces applied in two dimensions in a way that helps students to link those forces to an object's motion.

ThinkerTools is but one example of this type of interactive, representational software. Others have been developed to help students reason about motion (Roschelle, 1992), electricity (Gutwill, Fredericksen, and White, 1999), heat and temperature (Linn, Bell, and Hsi, 1998), genetics (Horwitz and Christie, 2000), and chemical reactions (Kozma, 2003), among others. These programs differ substantially from one another in how they represent their target phenomena, as there are substantial differences in the topics themselves and in the problems that students are known to have in understanding them. They share, however, a common approach to solving a similar set of problems—how to represent natural phenomena that are otherwise invisible in ways that help students make their own thinking explicit and guide them to normative scientific understanding.

When used as a supplement to hands-on laboratory experiences within integrated instructional units, these representations can support students' conceptual change (e.g., Linn et al., 1998; White and Frederiksen, 1998). For example, students working through the ThinkerTools curriculum always experiment with objects in the real world before they work with the computer tools. The goals of the laboratory experiences are to provide some experience with the phenomena under study and some initial ideas that can then be explored on the computer.

Structured Simulations of Inaccessible Phenomena

Various types of simulations of phenomena represent another form of technology for science learning. These simulations allow students to explore and observe phenomena that are too expensive, infeasible, or even dangerous to interact with directly. Strictly speaking, a computer simulation is a program that simulates a particular phenomenon by running a computational model whose behavior can sometimes be changed by modifying input parameters to the model. For example, the GenScope program provides a set of linked representations of genetics and genetics phenomena that would otherwise be unavailable for study to most students (Horowitz and Christie, 2000). The software represents alleles, chromosomes, family pedigrees, and the like and links representations across levels in ways that enable students to trace inherited traits to specific genetic differences. The software uses an underlying Mendelian model of genetic inheritance to govern its behavior. As with the representations described above, embedding the use of the software in a carefully thought out curriculum sequence is crucial to supporting student learning (Hickey et al., 2000).

Another example in biology is the BGuILE project (Reiser et al., 2001). The investigators created a series of structured simulations allowing students to investigate problems of evolution by natural selection. In the Galapagos finch environment, for example, students can examine a carefully selected set of data from the island of Daphne Major to explain a historical case of natural selection. The BGuILE software does not, strictly speaking, consist of simulations because it does not "run" a model; from a student's perspective, it simulates either Daphne Major or laboratory experiments on tuberculosis bacteria. Studies show that students can learn from the BGuILE environments when these environments are embedded in a well-organized curriculum (Sandoval and Reiser, 2004). They also show that successful implementation of such technology-supported curricula relies heavily on teachers (Tabak, 2004).

Structured Interactions with Complex Phenomena and Ideas

The examples discussed here share a crucial feature. The representations built into the software and the interface tools provided for learners are intended to help them learn in very specific ways. There are a great number of such tools that have been developed over the last quarter of a century. Many of them have been shown to produce impressive learning gains for students at the secondary level. Besides the ones mentioned, other tools are designed to structure specific scientific reasoning skills, such as prediction (Friedler et al., 1990) and the coordination of claims with evidence (Bell and Linn, 2000; Sandoval, 2003). Most of these efforts integrate students' work on the computer with more direct laboratory experiences. Rather than thinking of these representations and simulations as a way to replace laboratory experiences, the most successful instructional sequences integrate them with a series of empirical laboratory investigations. These sequences of science instruction focus students' attention on developing a shared interpretation of both the representations and the real laboratory experiences in small groups (Bell, 2005).

Computer Technologies Designed to Support Science

Advances in computer technologies have had a tremendous impact on how science is done and on what scientists can study. These changes are vast, and summarizing them is well beyond the scope of the committee's charge. We found, however, that some innovations in scientific practice, especially uses of the Internet, are beginning to be applied to secondary science education. With respect to future laboratory experiences, perhaps the most significant advance in many scientific fields is the aggregation of large, varied data sets into Internet-accessible databases. These databases are most commonly built for specific scientific communities, but some researchers are creating and studying new, learner-centered interfaces to allow access by teachers and schools. These research projects build on instructional design principles illuminated by the integrated instructional units discussed above.

One example is the Center for Embedded Networked Sensing (CENS), a National Science Foundation Science and Technology Center investigating the development and deployment of large-scale sensor networks embedded in physical environments. CENS is currently working on ecosystem monitoring, seismology, contaminant flow transport, and marine microbiology. As sensor networks come on line, making data available, science educators at the center are developing middle school curricula that include web-based tools to enable students to explore the same data sets that the professional scientists are exploring (Pea, Mills, and Takeuchi, 2004).

The interfaces professional scientists use to access such databases tend to be too inflexible and technical for students to use successfully (Bell, 2005). Bounding the space of possible data under consideration, supporting appropriate considerations of theory, and promoting understanding of the norms used in the visualization can help support students in developing a shared understanding of the data. With such support, students can develop both conceptual understanding and understanding of the data analysis process. Focusing students on causal explanation and argumentation based on the data analysis process can help them move from a descriptive, phenomenological view of science to one that considers theoretical issues of cause (Bell, 2005).

Further research and evaluation of the educational benefit of student interaction with large scientific databases are absolutely necessary. Still, the development of such efforts will certainly expand over time, and, as they change notions of what it means to conduct scientific experiments, they are also likely to change what it means to conduct a school laboratory.

SUMMARY

The committee identified a number of science learning goals that have been attributed to laboratory experiences. Our review of the evidence on attainment of these goals revealed a recent shift in research, reflecting some movement in laboratory instruction. Historically, laboratory experiences have been disconnected from the flow of classroom science lessons. We refer to these separate laboratory experiences as typical laboratory experiences. Reflecting this separation, researchers often engaged students in one or two experiments or other science activities and then conducted assessments to determine whether their understanding of the science concept underlying the activity had increased. Some studies compared the outcomes of these separate laboratory experiences with the outcomes of other forms of science instruction, such as lectures or discussions.

Over the past 10 years, researchers studying laboratory education have shifted their focus. Drawing on principles of learning derived from the cognitive sciences, they have asked *how* to sequence science instruction, including laboratory experiences, in order to support students' science learning. We refer to these instructional sequences as "integrated instructional units." Integrated instructional units connect laboratory experiences with other types of science learning activities, including lectures, reading, and discussion. Students are engaged in framing research questions, making observations, designing and executing experiments, gathering and analyzing data, and constructing scientific arguments and explanations.

The two bodies of research on typical laboratory experiences and integrated instructional units, including laboratory experiences, yield different findings about the effectiveness of laboratory experiences in advancing the science learning goals identified by the committee. The earlier research on typical laboratory experiences is weak and fragmented, making it difficult to draw precise conclusions. The weight of the evidence from research focused on the goals of developing scientific reasoning and enhancing student interest in science showed slight improvements in both after students participated in typical laboratory experiences. Research focused on the goal of student mastery of subject matter indicates that typical laboratory experiences are no more or less effective than other forms of science instruction (such as reading, lectures, or discussion).

Studies conducted to date on integrated instructional units indicate that the laboratory experiences, together with the other forms of instruction included in these units, show greater effectiveness for these same three goals (compared with students who received more traditional forms of science instruction): improving students' mastery of subject matter, increasing development of scientific reasoning, and enhancing interest in science. Integrated instructional units also appear to be effective in helping diverse groups of students progress toward these three learning goals. A major limitation of the research on integrated instructional units, however, is that most of the units have been used in small numbers of science classrooms. Only a few studies have addressed the challenge of implementing—and studying the effectiveness of—integrated instructional units on a wide scale.

Due to a lack of available studies, the committee was unable to draw conclusions about the extent to which either typical laboratory experiences or integrated instructional units might advance the other goals identified at the beginning of this chapter—enhancing understanding of the complexity and ambiguity of empirical work, acquiring practical skills, and developing teamwork skills. Further research is needed to clarify how laboratory experiences might be designed to promote attainment of these goals.

The committee considers the evidence sufficient to identify four general principles that can help laboratory experiences achieve the learning goals we have outlined. Laboratory experiences are more likely to achieve their intended learning goals if (1) they are designed with clear learning outcomes in mind, (2) they are thoughtfully sequenced into the flow of classroom science instruction, (3) they are designed to integrate learning of science content with learning about the processes of science, and (4) they incorporate ongoing student reflection and discussion.

Computer software and the Internet have enabled development of several tools that can support students' science learning, including representations of complex phenomena, simulations, and student interaction with large scientific databases. Representations and simulations are most successful in supporting student learning when they are integrated in an instructional sequence that also includes laboratory experiences. Researchers are currently developing tools to support student interaction with—and learning from—large scientific databases.

REFERENCES

Anderson, R.O. (1976). *The experience of science: A new perspective for laboratory teaching*. New York: Columbia University, Teachers College Press.

Ato, T., and Wilkinson, W. (1986). Relationships between the availability and use of science equipment and attitudes to both science and sources of scientific information in Benue State, Nigeria. *Research in Science and Technological Education, 4*, 19-28.

Beasley, W.F. (1985). Improving student laboratory performance: How much practice makes perfect? *Science Education, 69*, 567-576.

Bell, P. (2005). *The school science laboratory: Considerations of learning, technology, and scientific practice*. Paper prepared for the Committee on High School Science Laboratories: Role and Vision. Available at: http://www7.nationalacademies.org/bose/July_12-13_2004_High_School_Labs_ Meeting_Agenda.html [accessed June 2005].

Bell, P., and Linn, M.C. (2000). Scientific arguments as learning artifacts: Designing for learning from the web with KIE. *International Journal of Science Education, 22*(8), 797-817.

Ben-Zvi, R., Hofstein, A., Kampa, R.F, and Samuel, D. (1976). The effectiveness of filmed experiments in high school chemical education. *Journal of Chemical Education, 53*, 518-520.

Blakeslee, T., Bronstein, L., Chapin, M., Hesbitt, D., Peek, Y., Thiele, E., and Vellanti, J. (1993). *Chemistry that applies*. Lansing: Michigan Department of Education. Available at: http://www.ed-web2.educ.msu.edu/CCMS/secmod/Cluster3.pdf [accessed Feb. 2005].

Bransford, J.D., and Schwartz, D.L. (2001). Rethinking transfer: A simple proposal with multiple implications. In A. Iran-Nejad, and P.D. Pearson (Eds.), *Review of research in education* (pp. 61-100). Washington, DC: American Educational Research Association.

Bryce, T.G.K., and Robertson, I.J. (1985). What can they do: A review of practical assessment in science. *Studies in Science Education, 12,* 1-24.

Carey, S., and Smith, C. (1993). On understanding the nature of scientific knowledge. *Educational Psychologist, 28,* 235-251.

Champagne, A.B., Gunstone, R.F., and Klopfer, L.E. (1985). Instructional consequences of students' knowledge about physical phenomena. In L.H.T. West and A.L. Pines (Eds.), *Cognitive structure and conceptual change* (pp. 61-68). New York: Academic Press.

Chang, H.P., and Lederman, N.G. (1994). The effect of levels of co-operation within physical science laboratory groups on physical science achievement. *Journal of Research in Science Teaching, 31,* 167-181.

Cobb, P., Confrey, J., diSessa, A., Lehrer, R., and Schauble, L. (2003). Design experiments in educational research. *Educational Researcher, 32*(1), 9-13.

Cobb, P., Stephan, M., McClain, K., and Gavemeijer, K. (2001). Participating in classroom mathematical practices. *Journal of the Learning Sciences, 10,* 113-164.

Coleman, E.B. (1998). Using explanatory knowledge during collaborative problem solving in science. *Journal of the Learning Sciences, 7*(3, 4), 387-427.

Collins, A., Joseph, D., and Bielaczyc, K. (2004). Design research: Theoretical and methodological issues. *Journal of the Learning Sciences, 13*(1), 15-42.

Coulter, J.C. (1966). The effectiveness of inductive laboratory demonstration and deductive laboratory in biology. *Journal of Research in Science Teaching, 4,* 185-186.

Denny, M., and Chennell, F. (1986). Exploring pupils' views and feelings about their school science practicals: Use of letter-writing and drawing exercises. *Educational Studies, 12,* 73-86.

Design-Based Research Collective. (2003). Design-based research: An emerging paradigm for educational inquiry. *Educational Researcher, 32*(1), 5-8.

Driver, R. (1995). Constructivist approaches to science teaching. In L.P. Steffe and J. Gale (Eds.), *Constructivism in education* (pp. 385-400). Hillsdale, NJ: Lawrence Erlbaum.

Driver, R., Leach, J., Millar, R., and Scott, P. (1996). *Young people's images of science.* Buckingham, UK: Open University Press.

Driver, R., Newton, P., and Osborne, J. (2000). Establishing the norms of scientific argumentation in classrooms. *Science Education, 84,* 287-312.

Dunbar, K. (1993). Concept discovery in a scientific domain. *Cognitive Science, 17,* 397-434.

Dupin, J.J., and Joshua, S. (1987). Analogies and "modeling analogies" in teaching: Some examples in basic electricity. *Science Education, 73,* 791-806.

Duschl, R.A. (2004). *The HS lab experience: Reconsidering the role of evidence, explanation and the language of science.* Paper prepared for the Committee on High School Science Laboratories: Role and Vision, July 12-13, National Research Council, Washington, DC. Available at: http://www7.nationalacademies.org/bose/July_12-13_2004_High_School_Labs_ Meeting_Agenda.html [accessed July 2005].

Duschl, R.A., and Osborne, J. (2002). Supporting and promoting argumentation discourse in science education. *Studies in Science Education, 38,* 39-72.

Eylon, B., and Linn, M.C. (1988). Learning and instruction: An examination of four research perspectives in science education. *Review of Educational Research, 58*(3), 251-301.

Fraser, B.J., Giddings, G.J., and McRobbie, C.J. (1995). Evolution and validation of a personal form of an instrument for assessing science laboratory classroom environments. *Journal of Research in Science Teaching, 32,* 399-422.

Fraser, B.J., McRobbie, C.J., and Giddings, G.J. (1993). Development and cross-national validation of a laboratory classroom environment instrument for senior high school science. *Science Education, 77,* 1-24.

Freedman, M.P. (2002). The influence of laboratory instruction on science achievement and attitude toward science across gender differences. *Journal of Women and Minorities in Science and Engineering, 8,* 191-200.

Friedler, Y., Nachmias, R., and Linn, M.C. (1990). Learning scientific reasoning skills in microcomputer-based laboratories. *Journal of Research in Science Teaching, 27*(2), 173-192.

Glaser, R. (1994). Learning theory and instruction. In G. d'Ydewalle, P. Eelen, and P. Bertelson (Eds.), *International perspectives on science, volume 2: The state of the art* (pp. 341-357). Hove, England: Erlbaum.

Gobert, J., and Clement, J. (1999). The effects of student-generated diagrams versus student-generated summaries on conceptual understanding of spatial, causal, and dynamic knowledge in plate tectonics. *Journal of Research in Science Teaching, 36*(1), 39-53.

Gutwill, J.P., Fredericksen, J.R., and White, B.Y. (1999). Making their own connections: Students' understanding of multiple models in basic electricity. *Cognition and Instruction, 17*(3), 249-282.

Hatano, G., and Inagaki, K. (1991). Sharing cognition through collective comprehension activity. In L.B. Resnick, J.M. Levine, and S.D. Teasley (Eds.), *Perspectives on socially shared cognition* (pp. 331-348). Washington, DC: American Psychological Association.

Henderson, D., Fisher, D., and Fraser, B. (2000). Interpersonal behavior, laboratory learning environments, and student outcomes in senior biology classes. *Journal of Research in Science Teaching, 37,* 26-43.

Hickey, D.T., Kindfield, A.C.H., Horwitz, P., and Christie, M.A. (2000). Integrating instruction, assessment, and evaluation in a technology-based genetics environment: The GenScope follow-up study. In B.J. Fishman and S.F. O'Connor-Divelbiss (Eds.), *Proceedings of the International Conference of the Learning Sciences* (pp. 6-13). Mahwah, NJ: Lawrence Erlbaum.

Hickey, D.T., Kindfield, A.C., Horwitz, P., and Christie, M.A. (2003). Integrating curriculum, instruction, assessment, and evaluation in a technology-supported genetics environment. *American Educational Research Journal, 40*(2), 495-538.

Hodson, D. (1993). Philosophic stance of secondary school science teachers, curriculum experiences, and children's understanding of science: Some preliminary findings. *Interchange, 24,* 41-52.

Hofstein, A., and Lunetta, V.N. (1982). The role of the laboratory in science teaching: Neglected aspects of research. *Review of Educational Research, 52*(2), 201-217.

Hofstein, A., and Lunetta, V.N. (2004). The laboratory in science education: Foundations for the twenty-first century. *Science Education, 88,* 28-54.

Holden, C. (1990). Animal rights activism threatens dissection. *Science, 25*, 751.

Horowitz, P. (1996). Linking models to data: Hypermodels for science education. *High School Journal, 79*(2), 148-156.

Horowitz, P., and Christie, M.A. (2000). Computer-based manipulatives for teaching scientific reasoning: An example. In M.J. Jacobson and R.B. Kozma (Eds.), *Innovations in science and mathematics education: Advanced designs for technologies of learning* (pp. 163-191). Mahwah, NJ: Lawrence Erlbaum.

Jovanovic, J., and King, S.S. (1998). Boys and girls in the performance-based science classroom: Who's doing the performing? *American Educational Research Journal, 35*(3), 477-496.

Kesidou, S., and Roseman, J. (2002). How well do middle school science programs measure up? Findings from Project 2061's curriculum review. *Journal of Research in Science Teaching, 39*(6), 522-549.

Khishfe, R., and Abd-El-Khalick, F. (2002). Influence of explicit and reflective versus implicit inquiry-oriented instruction on sixth graders' views of nature of science. *Journal of Research in Science Teaching, 39*(7), 551-578.

Klopfer, L.E. (1990). Learning scientific enquiry in the student laboratory. In E. Hegarty-Hazel (Ed.), *The student laboratory and the science curriculum* (pp. 95-118). London, England: Routledge.

Kozma, R.B. (2003). The material features of multiple representations and their cognitive and social affordances for science understanding. *Learning and Instruction, 13*, 205-226.

Kuhn, D., Schauble, L., and Garcia-Mila, M. (1992). Cross-domain development of scientific reasoning. *Cognition and Instruction, 9*(4), 285-327.

Lazarowitz, R., and Tamir, P. (1994). Research on using laboratory instruction in science. In D.L. Gabel (Ed.), *Handbook of research on science teaching and learning* (pp. 94-130). New York: Macmillan.

Lederman, N.G. (1992). Students' and teachers' conceptions of the nature of science: A review of the research. *Journal of Research in Science Teaching, 29*(4), 331-359.

Lederman, N.G., Abd-El-Khalick, F., Bell, R.L., and Schwartz, R.S. (2002). Views of nature of science questionnaire: Toward valid and meaningful assessment of learners' conceptions of nature of science. *Journal of Research in Science Teaching, 39*(6), 497-521.

Lehrer, R., and Schauble, L. (2004). Scientific thinking and science literacy: Supporting development in learning contexts. In W. Damon, R. Lerner, K. Anne Renninger, and E. Sigel (Eds.), *Handbook of child psychology, sixth edition, volume four: Child psychology in practice*. Hoboken, NJ: John Wiley & Sons.

Lehrer, R., Schauble, L., Strom, D., and Pligge, M. (2001). Similarity of form and substance: Modeling material kind. In S.M. Carver and D. Klahr (Eds.), *Cognition and instruction: Twenty-five years of progress*. Mahwah, NJ: Lawrence Erlbaum.

Lemke, J. (1990). *Talking science: Language, learning, and values*. Norwood, NJ: Ablex.

Linn, M.C. (1997). The role of the laboratory in science learning. *Elementary School Journal, 97*, 401-417.

Linn, M.C. (2004). *High school science laboratories: How can technology contribute?* Presentation to the Committee on High School Science Laboratories: Role and Vision. June. Available at: http://www7.nationalacademies.org/bose/June_3-4_2004_High_School_Labs_Meeting_Agenda.html [accessed April 2005].

Linn, M.C., Bell, P., and Hsi, S. (1998). Using the Internet to enhance student understanding of science: The knowledge integration environment. *Interactive Learning Environments, 6*(1-2), 4-38.

Linn, M.C., Davis, E., and Bell, P. (2004a). Inquiry and technology. In M.C. Linn, E. Davis, and P. Bell, (Eds.), *Internet environments for science education.* Mahwah, NJ: Lawrence Erlbaum.

Linn, M.C., Davis, E., and Bell, P. (Eds.). (2004b). *Internet environments for science education.* Mahwah, NJ: Lawrence Erlbaum.

Linn, M.C., and Hsi, S. (2000). *Computers, teachers, peers.* Mahwah, NJ: Lawrence Erlbaum.

Linn, M.C., and Songer, B. (1991). Teaching thermodynamics to middle school children: What are appropriate cognitive demands? *Journal of Research in Science Teaching, 28*(10), 885-918.

Lunetta, V.N. (1998). The school science laboratory. In B.J. Fraser and K.G. Tobin (Eds.), *International handbook of science education* (pp. 249-262). London, England: Kluwer Academic.

Lynch, S. (2004). *What are the effects of highly rated, lab-based curriculum materials on diverse learners?* Presentation to the Committee on High School Science Laboratories: Role and Vision. July 12. Available at: http://www7.nationalacademies.org/bose/July_12-13_2004_High_School_Labs_Meeting_Agenda.html [accessed Oct. 2004].

Lynch, S., Kuipers, J., Pyke, C., and Szesze, M. (In press). Examining the effects of a highly rated science curriculum unitinstructional unit on diverse populations: Results from a planning grant. *Journal of Research in Science Teaching.*

Lynch, S., and O'Donnell, C. (2005). *The evolving definition, measurement, and conceptualization of fidelity of implementation in scale-up of highly rated science curriculum unitsintegrated instructional units in diverse middle schools.* Paper presented at the annual meeting of the American Educational Research Association, April 7, Montreal, Canada.

McRobbie, C.J., and Fraser, B.J. (1993). Associations between student outcomes and psychosocial science environment. *Journal of Educational Research, 87,* 78-85.

Meichtry, Y.J. (1993). The impact of science curricula on student views about the nature of science. *Journal of Research in Science Teaching, 30*(5), 429-443.

Metz, K.E. (2004). Children's understanding of scientific inquiry: Their conceptualization of uncertainty in investigations of their own design. *Cognition and Instruction, 22*(2), 219-290.

Meyer, K., and Woodruff, E. (1997). Consensually driven explanation in science teaching. *Science Education, 80,* 173-192.

Millar, R. (1998). Rhetoric and reality: What practical work in science education is really for. In J. Wellington (Ed.), *Practical work in school science: Which way now?* (pp. 16-31). London, England: Routledge.

Millar, R. (2004). *The role of practical work in the teaching and learning of science.* Paper prepared for the Committee on High School Science Laboratories: Role and Vision. Available at: http://www7.nationalacademies.org/bose/June3-4_2004_High_School_Labs_Meeting_Agenda.html [accessed April 2005].

National Research Council. (1999). *How people learn: Brain, mind, experience, and school.* Committee on Developments in the Science of Learning, J.D. Bransford, A.L. Brown, and R.R. Cocking (Eds.). Washington, DC: National Academy Press.

National Research Council. (2001). *Eager to learn: Educating our preschoolers.* Committee on Early Childhood Pedagogy. B.T. Bowman, M.S. Donovan, and M.S. Burns (Eds.). Commission on Behavioral and Social Sciences and Education. Washington, DC: National Academy Press.

National Research Council. (2005). *Systems for state science assessment.* Committee on Test Design for K-12 Science Achievement, M.R. Wilson and M.W. Bertenthal (Eds.). Board on Testing and Assessment, Center for Education. Division of Behavioral and Social Sciences and Education. Washington, DC: The National Academies Press.

Osborne, R., and Freyberg, P. (1985). *Learning in science: The implications of children's science.* London, England: Heinemann.

Partnership for 21st Century Skills. (2003). *Learning for the 21st century.* Washington, DC: Author. Available at: http://www.21stcenturyskills.org/reports/learning.asp [accessed April 2005].

Pea, R., Mills, M., and Takeuchi, L. (Eds). (2004). *Making SENS: Science education networks of sensors.* Report from an OMRON-sponsored workshop of the Media-X Program at Stanford University, October 3. Stanford, CA: Stanford Center for Innovations in Learning. Available at:: http://www.makingsens.stanford.edu/index.html [accessed May 2005].

Raghubir, K.P. (1979). The laboratory investigative approach to science instruction. *Journal of Research in Science Teaching, 16,* 13-18.

Reif, F., and St. John, M. (1979) Teaching physicists thinking skills in the laboratory. *American Journal of Physics, 47*(11), 950-957.

Reiner, M., Pea, R.D., and Shulman, D.J. (1995). Impact of simulator-based instruction on diagramming in geometrical optics by introductory physics students. *Journal of Science Education and Technology, 4*(3), 199-225.

Reiser, B.J., Tabak, I., Sandoval, W.A., Smith, B.K., Steinmuller, F., and Leone, A.J. (2001). BGuILE: Strategic and conceptual scaffolds for scientific inquiry in biology classrooms. In S.M. Carver and D. Klahr (Eds.), *Cognition and instruction: Twenty-five years of progress* (pp. 263-305). Mahwah, NJ: Lawrence Erlbaum.

Renner, J.W., Abraham, M.R., and Birnie, H.H. (1985). Secondary school students' beliefs about the physics laboratory, *Science Education, 69,* 649-63.

Roschelle, J. (1992). Learning by collaborating: Convergent conceptual change. *Journal of the Learning Sciences, 2*(3), 235-276.

Roschelle, J., Kaput, J., and Stroup, W. (2000). SimCalc: Accelerating students' engagement with the mathematics of change. In M.J. Jacobsen and R.B. Kozma (Eds). *Learning the sciences of the 21st century: Research, design, and implementing advanced technology learning environments* (pp. 47-75). Hillsdale, NJ: Lawrence Erlbaum.

Rosebery, A.S., Warren, B., and Conant, F.R. (1992). Appropriating scientific discourse: Findings from language minority classrooms. *Journal of the Learning Sciences, 2*(1), 61-94.

Salomon, G. (1996). Studying novel learning environments as patterns of change. In S. Vosniadou, E. De Corte, R. Glaser, and H. Mandl (Eds.), *International perspectives on the design of technology-supported learning environments* (pp. 363-377). Mahwah, NJ: Lawrence Erlbaum.

Sandoval, W.A. (2003). Conceptual and epistemic aspects of students' scientific explanations. *Journal of the Learning Sciences, 12*(1), 5-51.

Sandoval, W.A., and Millwood, K.A. (2005). The quality of students' use of evidence in written scientific explanations. *Cognition and Instruction, 23*(1), 23-55.

Sandoval, W.A., and Morrison, K. (2003). High school students' ideas about theories and theory change after a biological inquiry unit. *Journal of Research in Science Teaching, 40*(4), 369-392.

Sandoval, W.A., and Reiser, B.J. (2004). Explanation-driven inquiry: Integrating conceptual and epistemic supports for science inquiry. *Science Education, 88*, 345-372.

Schauble, L., Glaser, R., Duschl, R.A., Schulze, S., and John, J. (1995). Students' understanding of the objectives and procedures of experimentation in the science classroom. *Journal of the Learning Sciences, 4*(2), 131-166.

Schauble, L., Klopfer, L.E., and Raghavan, K. (1991). Students' transition from an engineering model to a science model of experimentation. *Journal of Research in Science Teaching, 28*(9), 859-882.

Shaffer, P.S., and McDermott, L.C. (1992). Research as a guide for curriculum development: An example from introductory electricity. Part II: Design of instructional strategies. *American Journal of Physics, 60*(11), 1003-1013.

Shepardson, D.P., and Pizzini, E.L. (1993). A comparison of student perceptions of science activities within three instructional approaches. *School Science and Mathematics, 93*, 127-131.

Shulman, L.S., and Tamir, P. (1973). Research on teaching in the natural sciences. In R.M.W. Travers (Ed.), *Second handbook of research on teaching*. Chicago: Rand-McNally.

Singer, R.N. (1977). To err or not to err: A question for the instruction of psychomotor skills. *Review of Educational Research, 47*, 479-489.

Smith, C.L., Maclin, D., Grosslight, L., and Davis, H. (1997). Teaching for understanding: A study of students' pre-instruction theories of matter and a comparison of the effectiveness of two approaches to teaching about matter and density. *Cognition and Instruction, 15*, 317-394.

Smith, C.L., Maclin, D., Houghton, C., and Hennessey, M. (2000). Sixth-grade students' epistemologies of science: The impact of school science experiences on epistemological development. *Cognition and Instruction, 18*, 349-422.

Snir, J., Smith, C.L., and Raz, G. (2003). Linking phenomena with competing underlying models: A software tool for introducing students to the particulate model of matter. *Science Education, 87*(6), 794-830.

Songer, N.B., and Linn, M.C. (1991). How do students' views of science influence knowledge integration? *Journal of Research in Science Teaching, 28*(9), 761-784.

Tabak, I. (2004). Synergy: a complement to emerging patterns of distributed scaffolding. *Journal of the Learning Sciences, 13*(3), 305-335.

Tasker, R. (1981). Children's views and classroom experiences. *Australian Science Teachers' Journal, 27*, 33-37.

Tiberghien, A., Veillard, L., Le Marechal, J.-F., Buty, C., and Millar, R. (2000). An analysis of labwork tasks used in science teaching at upper secondary school and university levels in several European countries. *Science Education, 85*, 483-508.

Tobin, K. (1987). Forces which shape the implemented curriculum in high school science and mathematics. *Teaching and Teacher Education, 3*(4), 287-298.

VandenBerg, E., Katu, N., and Lunetta, V.N. (1994). *The role of "experiments" in conceptual change.* Paper presented at the annual meeting of the National Association for Research in Science Teaching, Anaheim, CA.

Webb, N.M., Nemer, K.M., Chizhik, A.W., and Sugrue, B. (1998). Equity issues in collaborative group assessment: Group composition and performance. *American Educational Research Journal, 35*(4), 607-652.

Webb, N.M., and Palincsar, A.S. (1996). Group processes in the classroom. In D.C. Berliner and R.C. Calfee (Eds.), *Handbook of educational psychology* (pp. 841-873). New York: Macmillan.

Wells, M., Hestenes, D., and Swackhamer, G. (1995). A modeling method for high school physics instruction. *American Journal of Physics, 63*(7), 606-619.

Wheatley, J.H. (1975).Evaluating cognitive learning in the college science laboratory. *Journal of Research in Science Teaching, 12*, 101-109.

White, B.Y. (1993). ThinkerTools: Causal models, conceptual change, and science education. *Cognition and Instruction, 10*(1), 1-100.

White, B.Y., and Frederiksen, J.R. (1998). Inquiry, modeling, and metacognition: Making science accessible to all students. *Cognition and Instruction, 16*(1), 3-118.

White, R.T. (1996). The link between the laboratory and learning. *International Journal of Science Education, 18*, 761-774.

White, R.T., and Gunstone, R.F. (1992). *Probing understanding.* London, England: Falmer.

Wilkenson, J.W., and Ward, M. (1997). The purpose and perceived effectiveness of laboratory work in secondary schools. *Australian Science Teachers' Journal*, 43-55.

Wong, A.F.L., and Fraser, B.J. (1995). Cross-validation in Singapore of the science laboratory environment inventory. *Psychological Reports, 76*, 907-911.

Woolnough, B.E. (1983). Exercises, investigations and experiences. *Physics Education, 18*, 60-63.

Yager, R.E., Engen, J.B., and Snider, C.F. (1969). Effects of the laboratory and demonstration method upon the outcomes of instruction in secondary biology. *Journal of Research in Science Teaching, 5*, 76-86.

Zimmerman, C. (2000). The development of scientific reasoning skills. *Developmental Review, 20*, 99-149.

4

Current Laboratory Experiences

Key Points

- *Laboratory experiences have unique features that pose a challenge to effective instruction.*
- *On average, high school students enrolled in science classes participate in laboratory experiences for one class period each week; however, students in schools with higher concentrations of non-Asian minorities spend less time in laboratory experiences than students in schools with fewer non-Asian minorities.*
- *Most high school students participate in typical laboratory experiences that are isolated from the flow of science instruction and do not follow the other design principles outlined by the committee.*

The previous chapter reviewed research on the outcomes of different types of laboratory experiences and outlined principles of instructional design to guide development of more effective laboratory experiences. In this chapter, the committee reviews evidence about the quantity and quality of laboratory experiences in U.S. high schools today. We begin with a description of the nature of laboratory education, which poses challenges to teachers and schools, and then address how these challenges are being met. The next section focuses on the amount of time science students spend in laboratory activities as part of their science courses. We then assess current laboratory experiences in light of the range of experiences presented in Chapter 1 and the goals and instructional design principles presented in Chapter 3. The chapter concludes that most laboratory experiences today are "typical" laboratory experiences, isolated from the flow of science instruction. Because these typical laboratory experiences do not follow the design principles we have outlined, they are unlikely to help students attain the science learning goals identified in Chapter 3:

- Enhancing mastery of subject matter.
- Developing scientific reasoning.
- Understanding the complexity and ambiguity of empirical work.
- Developing practical skills.
- Understanding of the nature of science.
- Cultivating interest in science and interest in learning science.
- Developing teamwork abilities.

THE UNIQUE NATURE OF LABORATORY EXPERIENCES

Laboratory experiences have features that make them unlike other forms of science instruction. These unique features make it a challenge to structure laboratory experiences so that they neither overwhelm students with complexity on one hand nor rigidly specify all of the questions, procedures, and materials on the other. Over the course of a student's high school science career, the appropriate balance between complexity and specificity may vary.

Students' direct interactions with the material world are inherently ambiguous, complex, and messy. Other modes of science instruction, such as lectures, readings, and homework problems, present students with simplified representations of natural phenomena that select and communicate certain variables and attributes (Millar, 2004). Although this simplification is essential for effective learning, it can create distance between classroom learning and real-world applications of science. Students may find that a problem-solving approach that worked well in the classroom fails badly when applied to observation or manipulation of the material world.

Natural phenomena contain much more information than any representation (Millar, 2004), and this wealth of information and complexity can prevent students and teachers from focusing on and attaining the goals of laboratories we have outlined. For example, when discussing a pendulum in class, a physics teacher may ignore without discussion a host of variables that may affect its operation. However, when a student starts doing a simple experiment with a pendulum, these variables suddenly become relevant.

Relevant variables begin with physical forces, including friction and air resistance, and continue through the range of complexity to the air pressure and wind in the room, and they include limits on human reaction time and the acuity of a student's vision. Along with the enormous increase in the number of possible relevant variables comes the problem of sorting out which ones matter and which do not. This problem can quickly become overwhelming to the student and the instructor.

A student can become frustrated and confused when almost everything seems to matter in a variety of mysterious ways. Should she or he worry about the amount of sound in the room, how warm it is, and whether it is in a basement or on the third floor? The student may feel betrayed by the apparent mismatch between the neatness of a phenomenon as presented in a textbook and the inherent messiness and ambiguity of the same phenomenon encountered in the laboratory. The instructor is similarly confronted with a host of complexities that put enormous demands on both his or her knowledge of the material (and experimental science in general) and ability to turn the student's confusion and frustration into an educationally valuable experience.

In addition to these problems of frustration and confusion, students sometimes make observations or gather data during laboratory experiences that appear to contradict known scientific principles or concepts (Olsen, Hewson, and Lyons, 1996; Hammer, 1997). To avoid this and to keep students from being overwhelmed by complexity, laboratory manuals and teachers may constrain the types of questions studied and the procedures used to answer these questions (Olsen et al., 1996). Schools and teachers may also respond to these challenges by scheduling fewer or shorter laboratory activities (or eliminating them entirely). The following sections describe current responses to the unique nature of laboratory experiences.

QUANTITY OF LABORATORY INSTRUCTION

The amount of time high school students spend in laboratory experiences is related to the number and level of science courses they take and to the demographics of the school. Students in science courses generally taken after introductory biology and students in schools with fewer non-Asian minorities generally spend more time in laboratory instruction than do students in other science courses and students in schools with high concentrations of non-Asian minorities.

Science Courses and Laboratory Experiences

Science Course-Taking

Over the past 20 years, the percentage of high school graduates taking more than two years of science classes has grown. In 1982, high school graduates earned an average of 2.2 science credits (1 credit equals 1 year of a course that meets daily). By 1998, the number grew to 3.2 credits. This expansion in the number of science courses taken included all racial/ethnic groups and both male and female students (National Center for Education Statistics, 2004).

The proportion of high school graduates taking science courses after completing general biology has also grown. From 1982 to 2000, the percentage of high school graduates who had completed at least one course beyond general biology increased from 35 to 63 percent, primarily because more students completed introductory chemistry or physics (or both). In 2000, most high school graduates (63 percent) had completed at least one class after taking general biology, 30 percent completed either chemistry or physics, and about 18 percent had completed the highest level classes, which are equivalent to introductory college science courses—advanced placement (AP) and international baccalaureate (IB) biology, chemistry, and physics classes (National Center for Education Statistics, 2004). Other surveys, conducted in conjunction with the National Assessment of Educational Progress (NAEP), found that the fraction of high school graduates who had completed AP or IB science courses increased from 7.8 percent in 1998 to 9.1 percent in 2000 (Perkins, Kleiner, Roey, and Brown, 2004).

Surveys conducted in conjunction with NAEP also indicate that most students do not take four full years of high school science. In response to this survey, 53 percent of 12th graders indicated they were enrolled in a science course. The 12th grade students indicated that most had taken biology in 9th or 10th grade, but fewer had completed chemistry and physics courses in 11th or 12th grade (O'Sullivan, Lauko, Grigg, Quian, and Zhang, 2003).

Features of Current Laboratory Experiences

The available data indicate that the average high school student takes science classes during three of the four years of high school and participates in laboratory activities approximately once a week during these science classes.

Horizon Research, Inc. conducted national surveys of science and mathematics education for the National Science Foundation (NSF) in 1977, 1982, 1993, and 2000 (Smith, Banilower, McMahon, and Weiss, 2002). The surveys included probability samples of schools and teachers, designed to yield nationally representative results, and received high response rates from teachers and principals. Among other questions, the surveys asked about instructional practices (survey results related to laboratory facilities are discussed in Chapter 6).

In response to the year 2000 survey, 71 percent of high school teachers reported that they involved students in "hands-on/laboratory science activities or investigations," at least once a week, representing a small but statistically significant increase from the 67 percent of high school teachers who reported such activities in 1993. In both 1993 and 2000, high school teachers reported using about 20 percent of instructional time (about one day a week) for laboratory activities (Smith et al., 2002).

Survey data indicate that many current laboratory experiences are restricted in their settings and use of technology. For example, 50 percent of teachers responding to the 2000 survey indicated that they never took field trips. When the researchers compared national survey results in 1993 and 2000, they found (Smith et al., 2002, p. 43): "the use of computers in science lessons is striking in its lack of change. Even in 2000, less than 10 percent of science lessons included students using computers."

In 2000, 45 percent of science teachers indicated that they never used laboratory simulations, 54 percent never engaged students in solving problems using simulations, 55 percent never engaged students in collecting data using sensors or probes, and 43 percent never engaged students in retrieving or exchanging data over the Internet. Data collected as part of the NAEP science assessment revealed similarly low levels of technology use in science classrooms (O'Sullivan et al., 2003). It appears that there is a considerable gap between the potential of computer technology to aid student learning in laboratory experiences discussed in Chapter 3 and the current reality.

Disparities in Laboratory Experiences

Variation in Course-Taking

There are racial/ethnic differences in enrollment in the advanced science courses that include more minutes of laboratory instruction. A study of student participation in science courses between 1982 and 1992 found that, at both points in time, black and Hispanic students took fewer science courses than white or Asian students (Quinn, 1996). During most of the 1990s, as high schools offered more science courses, an increasing proportion of students took more advanced courses, but racial/ethnic differences persisted. By 2000, Asians were more likely than students of any other ethnicity to have completed chemistry, physics, and other science courses usually taken after completing general biology, but there was no statistically significant difference in the percentage of white, black, and Hispanic high school graduates who had completed such courses (National Center for Education Statistics, 2004). High school graduates from urban and suburban schools were generally more likely than their counterparts from rural schools to have completed science courses beyond general biology. Participation in AP/IB biology and AP/IB chemistry increased with school size. In addition, as school poverty increased, fewer students completed courses in chemistry and physics (National Science Foundation, 2004).

Data on ethnic group participation in the AP examinations (which most students who enroll in AP courses take) provide indirect evidence of disparities in enrollment in AP courses. The College Board recently found that white test-takers were roughly proportionate to their representation in the school population as a whole (67.5 percent of the school population and 64.5 percent of AP test-takers). The same was true of Hispanic students, who made up 12.8 percent of the 2004 school population and 13.1 percent of those who took the AP test in 2004. However, black students made up 13.2 percent of the school population but only 6 percent of students who took one or more AP exams in 2004, and American Indians made up 1.1 percent of the school population but only 0.5 percent of AP test-takers. In contrast to these ethnic groups, a disproportionately large share of Asian students took the AP exam (10.6 percent of all AP test-takers) in comparison to their fraction of the total school population (5.1 percent) (College Board, 2005). These data describe the population that took any type of AP test and are not specific to those who took AP science tests.

Science Course Offerings

Variations in patterns of course-taking, especially among poor and minority students, may reflect differences in the kinds of courses offered in schools with different populations of students. For example, one study found that black students enroll in fewer physical science courses, and schools with larger black student populations are likely to offer fewer physical science course opportunities (Norman et al., 2001).

Data on science course offerings gathered in 1990, 1994, and 1998 in conjunction with NAEP show course offerings that vary with the characteristics of schools and students. These data indicate that chemistry, physics, and other science courses usually taken after completing general biology science courses were widely available (90 percent of graduates in all three years attended high schools that offered such courses), but AP and IB courses were less widespread. In 1998, schools attended by 46 percent of graduates offered AP/IB biology, schools attended by 39 percent of graduates offered AP/IB chemistry, and only 27 percent of graduates attended schools offering AP/IB physics (National Science Foundation, 2004). When NSF staff analyzed the survey data, they found that urban and suburban schools, as well as larger schools, more frequently offered advanced science courses than rural schools and smaller schools. They also found that wealthier schools were much more likely to offer AP/IB chemistry and physics classes than schools with high percentages of poor students (National Science Foundation, 2004).

One study found that the availability of AP offerings in California varied with the school's racial and socioeconomic composition. The availability of AP courses decreased as the percentage of black, Hispanic, or poor students in the school population increased (Oakes et al., 2000).

Although the problem of uneven participation in advanced science courses (and in the laboratory experiences they provide) may be partly explained by inequities in courses offered, other factors are also important. When schools do offer advanced science courses, minority and low-income students are much less likely than other students to enroll in them. (Atanda, 1999; Oakes, 1990; Oakes, Gamoran, and Page, 1992).

Disparities in Laboratory Experiences

Several sources of evidence indicate that the amount of time students spend in laboratory experiences varies based on their ethnicity and level of science courses taken.

A follow-up analysis of data from the 2000 survey of science teachers and schools described above revealed disparities in the frequency and duration of laboratory experiences (Banilower, Green, and Smith, 2004). The authors analyzed data on time spent in various types of science instruction in general and on time spent in various forms of science instruction in the most recent science lesson. For purposes of the study, they grouped the schools included in the survey into four levels of concentration of non-Asian minorities. They found that, during the most recent science lesson, students in schools with the fewest non-Asian minority students spent significantly more time "working with hands-on, manipulative, or laboratory materials" than students in schools with the highest concentration of non-Asian minority students (Banilower et al., 2004, p. 30). They also found that teachers in schools with the second highest and highest concentrations of minority students were significantly more likely than teachers in other schools to engage students in individually reading texts or completing worksheets.

The National Education Longitudinal Study of 1988 analyzed data on high school seniors in 1992 (Quinn, 1996). This study found that the frequency of laboratory experiences varied according to the achievement level of the class (as reported by the teacher). On average, across all classes, 57 minutes per week were allocated for science laboratory activities. Among AP classes, an average of 76 minutes per week was allocated to laboratory activities per week. In low-achievement-level courses, 40 minutes per week was allocated to laboratory activities, compared with 50 minutes per week in average-level classes and 61 minutes for high-achievement-level classes. Other approaches to science instruction also varied by achievement level of the class. In AP courses, a greater percentage of time was spent in whole-class instruction (57 percent) compared with low-achievement classes (47 percent), and less time was spent maintaining order (2 percent of time versus 9 percent of time for low-achievement-level classes). Also, teachers were more likely to lecture in higher achievement level courses, to allow students to respond orally to questions on subject matter, and to use computers. Students in the higher achievement classes were less likely to complete individual written assignments or worksheets in class (Quinn, 1996).

In this longitudinal study, Quinn created regression models to explore the relation between socioeconomic status and science teachers' instructional strategies. When achievement level of the class was not taken into account, students with higher socioeconomic levels received more minutes of laboratory instruction per week. However, when achievement level of the class is taken into account, the effect of socioeconomic status disappeared. A similar effect was obtained for emphasis on inquiry (the processes of science) (Quinn, 1996).

QUALITY OF CURRENT LABORATORY EXPERIENCES

The extent to which current laboratory experiences help students attain educational goals depends not only on how many minutes are spent in laboratory instruction but also on the quality of that instruction.

Comparison with Instructional Design Principles

Research indicates that laboratory experiences are more likely to help students attain learning goals if they are:

- designed with clear outcomes in mind,
- sequenced into the flow of classroom science instruction,
- designed to integrate learning of science content and process, and
- incorporated for ongoing student reflection and discussion.

Lack of Focus on Clear Learning Goals

Today's high school laboratory experiences are not always designed with clear learning outcomes in mind. The effectiveness of a laboratory activity can be assessed in terms of outcomes at two different and interdependent levels, a basic level and the level of desired learning outcomes. In order to be effective in achieving its desired learning outcomes, a laboratory activity must first be effective at the basic level—the students must carry out the activities and obtain the results intended by the designer (Millar, 2004). The inherent complexity and ambiguity of laboratory activities may prevent students from achieving even basic effectiveness. In order to help ensure that they do indeed carry out the activities as intended, laboratory manuals and teachers often provide detailed procedures (Tobin and Gallagher 1987; De Carlo and Rubba, 1994; Priestley, Priestley, and Schmuckler, 1997; Millar,

2004). The resulting "cookbook" activity may reduce the possibility that students' observations and analysis will lead to conclusions that are at odds with accepted scientific principles, but it may also hamper effectiveness at the higher level (attainment of desired learning outcomes).

When curriculum developers or teachers focus on the goal of foolproof results, they are less likely to design or carry out the laboratory experience with clear learning goals in mind. And, as discussed in the previous chapter, when teachers and students are unclear about the learning goals of laboratory experiences, they are less likely to attain those goals. One experienced high school physics teacher found that several years of providing increasingly detailed instructions helped students in "doing the lab right" (Olsen et al., 1996, p. 785), but it did not help them develop any ideas about the purposes of the laboratory activities.

Isolated from the Flow of Science Instruction

Another problem is that many current laboratory experiences are not well integrated into the stream of instruction (Sutman, Schmuckler, Hilosky, Priestley, and Priestley, 1996). Laboratory activities often remain disconnected and isolated from instruction, rather than being explicitly integrated with lectures, reading, and discussion (Linn, Songer, and Eylon, 1996; Linn, 2003).

Even when laboratory activities are designed in ways that integrate at least partially into the stream of instruction and with clear learning goals in mind, they are not always implemented as planned. The AP Biology Lab Manual for Teachers (College Board, 2001) presents a sequence of five laboratory experiences focusing on diffusion and osmosis. Although this laboratory manual provides no guidance on how to integrate this series of experiences with other forms of instruction or previous biology topics covered, the laboratory experiences themselves are carefully sequenced. The initial two activities engage students in experimenting with dialysis tubing as a model of a cell membrane. In three later activities, students observe osmosis and diffusion in real plant cells, first in a potato core and then in red onion cells. Both this progression of activities and the laboratory manual itself clarify the learning goals for the series (College Board, 2001, p. 1):

1. investigate the processes of diffusion and osmosis in a model membrane system and
2. investigate the effect of solute concentration on water potential as it relates to living plant tissues.

These two objectives clearly state the underlying goal of helping students to understand the activity of plant membranes and cells. The lab manual suggests the amount of time needed to complete each activity; following the

BOX 4-1 Diffusion Across a Selectively Permeable Membrane

> The laboratory activity is intended to address if the size of a molecule affects whether or not it can diffuse across a selectively permeable membrane. To carry out the experiment, students fill dialysis tubing with a solution of glucose and starch. They place the dialysis tubing in a beaker that contains IKI, a color indicator for starch. Students determine if glucose moves out of the dialysis tubing into the beaker by dipping a test strip into the beaker and checking to see if it changes color (indicating the presence of glucose). Students simultaneously determine if starch crosses this selectively permeable membrane by observing if the color of the water in the beaker changes.
>
> SOURCE: Adapted from <http://www.sc2000.net/~czaremba/aplabs/osmosis.html> and the College Board (2001).

suggestions would require 6-7 periods of science class (assuming a 45-minute class period).

Teachers may not carry out this full sequence of osmosis and diffusion activities, for at least two reasons. First, when they see AP assessments as pressing them to cover multiple science topics, AP school biology teachers may choose to carry out only one or two activities, rather than devoting 6-7 class periods to these topics. Second, teachers and schools who lack funds to purchase the AP biology lab manual can find simplified versions of the first, or the first two, osmosis and diffusion laboratory activities on the Internet at no cost (http://www.ekcsk12.org/science/aplabreview/aplabonediffusionandosmosis.htm and http://www.sc2000.net/~czaremba/aplabs/osmosis.html). These two activities use dialysis tubing as a model of a cell membrane, but neither of the two Internet versions of the activities includes the two learning goals stated in the AP lab manual, which clarify the underlying goal of helping students to understand living plant tissues.

If the first laboratory activity is carried out in isolation from the sequence of other laboratory activities and in isolation from lectures, discussion, and other modes of learning, it may not help students progress in attaining laboratory learning goals (see Box 4-1).

In theory, this experience will enhance students' understanding of cell membranes and diffusion across cell membranes, which would help achieve one of the goals of laboratory experiences we have identified—enhancing mastery of subject matter. The activity could theoretically help students attain other goals of laboratory experiences, such as helping them develop scientific reasoning as they gather and interpret their data. Students may also gain an appreciation of the ambiguity and complexity of empirical work if they obtain conflicting results and through discussion are forced to reflect on and consider the sources of these discrepancies. The activity is not specifically designed to teach practical skills or help students develop an understanding of the nature of science.

In practice, this laboratory experience is unlikely to help students attain educational goals unless the teacher can integrate it into the stream of instruction. If the teacher embeds the experiment in instruction on selectively permeable membranes and cells, then it is more likely to help students master this subject matter. If the teacher clarifies the learning goal of the laboratory experience by presenting the dialysis tubing to the student as a model of a cell membrane, then substantially more biological subject matter learning may occur. Similarly, a teacher may decide to ask students to carry out the experiment in two steps, first testing for glucose diffusion and then for starch diffusion. This would eliminate the potential confusion created by testing both starch and glucose diffusion simultaneously. Finally, the teacher may encourage students to discuss among themselves what their results mean and follow up with a whole-class discussion. This opportunity for reflection might help to improve students' ability to interpret and make inferences from data. Although an expert teacher may focus on desired learning goals and integrate this activity into the curriculum in order to help students attain those goals, teachers often focus instead on the laboratory procedures themselves.

Little Integration of Science Content and Science Process

Current laboratory experiences do not always integrate the learning of science content with learning about the processes of science, perhaps because of the unique challenges presented by laboratory experiences. As noted at the beginning of this chapter, a real pendulum in a high school physics classroom brings with it a host of potentially confusing variables. To reduce the potential confusion and to help students attain one goal—mastery of subject matter—a typical high school pendulum activity is "cleaned up." This activity is designed to guide students toward making observations that will verify the accepted scientific principle that the period of a pendulum (the time it takes to swing out and back) depends on the length of the string and the force of gravity. It focuses only on science content. Laboratory manuals and teachers rarely use an alternative approach to a pendulum activity designed to help students understand not only this known principle but also the process scientists use to establish such principles (see Box 4-2).

Although both of these approaches may help students to develop skills in making observations and gathering and presenting data, only the second integrates the learning of science content and process. The first approach is designed to foster students' mastery of science subject matter, by verifying a known physical relationship. The second approach fosters other goals of the laboratory experience, including understanding the complexity and ambiguity of empirical work, developing scientific reasoning ability, and understanding the nature of science. The second approach is designed to achieve these science process goals in the context of an activity that verifies a known scientific principle, so that it may help students to simultaneously master science subject matter.

Lack of Reflection and Discussion

There is evidence that current laboratory activities rarely incorporate ongoing reflection and discussion, although such discussion can enhance the effectiveness of laboratory learning. Data from the 2000 survey of science teachers indicate that only a third of teachers ask students to present their work once or twice a month, while another 44 percent of teachers use student presentations only a few times each year, although such presentations often lead to discussion and reflection (Weiss, Banilower, McMahon, and Smith, 2001). The survey also indicates that teachers rarely engage students in small-group discussions. A study of laboratory experiences in three high school chemistry classrooms found that the teachers rarely asked the kinds of questions that might generate discussion and reflection on science concepts (DeCarlo and Rubba, 1994); see Box 4-3. In general, the research indicates that students have few opportunities to construct shared understanding of scientific concepts as part of a community of learners in the classroom (Lunetta, 1998).

When discussion does take place during typical laboratory experiences, teachers and students often focus on procedures rather than processes and concepts (Hegarty-Hazel, 1990, cited in Lazarowitz and Tamir, 1994). As a result, students have few opportunities to reflect on and develop their understanding of the scientific concepts underlying their laboratory experiences.

Comparison with a Range of Laboratory Experiences

Over the course of the high school years, a variety of laboratory experiences can help students to experience the range of activities that are part of the work of research scientists. Students' understanding of the processes of

BOX 4-2 Learning Physics Using a Pendulum: Two Approaches

> Most physics textbooks, laboratory manuals, and classrooms include a carefully limited cookbook pendulum activity. Each lab group is given a pendulum with the same mass and is asked to pull the pendulum back the same angle each time and see what happens to the period when the length of the string is changed. The students may even be told to vary the length of the string by increments of 10 cm before each pull. The students are provided with a data chart to fill in the length of the string and the period. The students are then asked to complete a graph of period versus length. When the charts and graphs are filled in, the students hand them in for grading by the teacher.
>
> A few physics texts, laboratory manuals, and teachers take a different approach. The pendulum activity in one curriculum (Hestenes et al., 2002) is designed to develop students' skills in designing experiments, collecting data, mathematical modeling, and reporting interpretations. Before the laboratory activity, the teacher demonstrates swinging pendulums with at least three different masses and engages students in observing and discussing the behavior of the pendulums. The teacher leads a prelab discussion, helping students identify variables that may affect the period, including variables that cannot be controlled (room temperature, gravity) and variables that can be controlled. Through discussion, the

science can be enhanced when laboratory experiences provide opportunities to:

- pose a research question,
- use laboratory tools and procedures,
- make observations, gather, and analyze data,
- verify, test, or evaluate explanatory models (including verifying known scientific theories and laws),
- formulate alternative hypotheses,
- design investigations, and
- build or revise explanatory models.

Few high school students today participate in this full range of laboratory experiences. In response to the 2000 survey of science teachers, 61 teacher helps students identify which controllable variables are related in order to isolate the dependent variable (period) and the independent variables (length, mass, and amplitude). Following a predict-observe-explain approach (see Chapter 3), the teacher asks students to make tentative predictions about how changes in the independent variables will affect the period.

After this discussion, in the laboratory activity, the teacher demonstrates various trials, pulling the pendulums of different masses back at different amplitudes and using different lengths of string. Each student is given a stopwatch to gather data on these trials, and one student records the observations on the blackboard. Students are asked to graph the relationships between period and mass, period and amplitude, and period and length. Following this laboratory activity, the teacher leads a discussion of the importance of adequate data quantity and ranges, modeling, the concepts of dependent and independent variables, and the definitions of period, frequency, weight, and mass. The teacher is told to avoid introducing the formal pendulum equation, because the laboratory activity is not designed to verify this known relationship.

SOURCE: Hestenes, Jackson, Dukerich, and Swackhamer (2002).

percent of high school teachers indicated that they engaged students in hands-on or laboratory science activities once or twice a week, and nearly the same fraction of teachers (59 percent) indicated that students followed specific instructions in an activity or investigation once or twice a week (Banilower et al., 2004). Students are likely to require specific instructions in order to use laboratory tools and procedures, make observations and gather data), while specific instructions would be more difficult to develop and apply to the more complex activities such as formulating a research question and designing an investigation. The frequent use of specific instructions may reflect an emphasis on laboratory experiences that focus on learning to use tools, make observations, and gather data.

Detailed observational case studies and National Research Council (NRC) studies also suggest that current laboratory experiences are primarily restricted to the first three types of opportunities listed earlier. Studies of

BOX 4-3 Examples of High School Chemistry Laboratory Experiences

Two researchers conducted a study of the behavior of three rural Pennsylvania high school chemistry teachers and their students during laboratory activities (DeCarlo and Rubba, 1994). They used a specially designed systematic observation instrument to code teacher and student behaviors during consecutive laboratory activities from November through April. They found that each teacher used a characteristic teaching style throughout this period.

Teacher 1 was highly interactive and social, circulating around the room and talking with students. These conversations usually focused on telling the students what to do or simply on socializing. Teacher 2 was somewhat interactive and somewhat unengaged, and this teacher's interactions with students were not related to the laboratory activities. Teacher 3 was unengaged, spending most of his time at his desk, where he graded papers, read journals, or did other tasks.

Students of Teacher 1 spent most of the laboratory period manipulating equipment and making observations, and their discussions focused on procedures rather than on interpreting results. Students of Teacher 2 spent most of their time socializing, although they also were engaged in fetching materials, manipulating equipment, and making observations during a few laboratory periods. Students of Teacher 3, like those of Teacher 1, were most frequently engaged in manipulating equipment and making observations. In comparison to students of the other two teachers, students of Teacher 3 were more frequently engaged in discussions related to their laboratory investigations and were less often unengaged in the laboratory activities.

The researchers found that none of the chemistry teachers ever asked questions aimed at encouraging students to think about what they were doing, even though all three had indicated that they did so frequently. In each case, the teachers responded to students' questions with specific answers, rather than by probing more deeply to understand the reason for the question. Another important finding was that the Teacher 3's lack of assistance forced the students to think and act on their own, a possibility identified in an earlier study by Shymansky and Penick (1981).

SOURCE: Adapted from DeCarlo and Rubba (1994).

teachers, students, and their interactions have found that teachers tend to emphasize specific instructions (which will help ensure that students' results verify known scientific principles) even when the teachers have stated that their goal is to stimulate student thinking (DeCarlo and Rubba, 1994; Marx, Freeman, Krajcik, and Blumenfeld, 1998). An NRC committee found that high school chemistry laboratory experiences tend more toward verification than problem-solving investigations (National Research Council, 2002, p. 356). Another committee found that laboratory exercises in AP biology courses "tend to be 'cookbook,' rather than inquiry based" (National Research Council, 2002, p. 292).[1] A recent review of the literature on laboratory education notes that "very often teachers involved students principally in relatively low-level, routine activities in laboratories" (Hofstein and Lunetta, 2004, p. 39).

Frequent laboratory activities emphasizing the use of scientific tools and procedures, gathering data, and verifying known concepts may leave little time for students to formulate research questions, analyze their data, or develop and revise models to explain the data. One study of 12 high school chemistry classes and 26 undergraduate chemistry classes found that there was rarely any follow-up discussion or analysis of data obtained during laboratory activities. In some cases, student groups were asked to combine their data with that of other groups, but the combined data were then never referred to again (Sutman et al., 1996). In response to the 2000 survey, 46 percent of teachers indicated that they asked students to record, represent, or analyze their data once or twice a week and 38 percent asked students to do so once or twice a month. Recording, representing, and analyzing data are essential steps in building and revising explanatory models, but it appears that many laboratory experiences do not include these activities.

In the 2000 survey, only 8 percent of high school teachers indicated that they asked students to design or implement their own investigation once or twice a week. Another 41 percent of teachers said students were asked to design or implement their own investigation once or twice a month; 42 percent of teachers indicated students were asked to design or implement their own investigation a few times a month. As a result, current laboratory experiences may provide few opportunities for students to make progress toward such goals as developing scientific reasoning abilities, understanding the complexity and ambiguity of empirical work, and understanding the nature of science.

---

[1]Since publication of this report, the College Board has changed AP tests and is developing technical assistance to teachers to encourage a wider range of laboratory experiences.

What Students Do in Laboratory Experiences

Only a few studies provide detailed descriptions of how students and teachers behave and interact during laboratory experiences. Among these, several focused on the differences between the behavior of boys and girls.

In a study summarizing results from several observational studies of science classes in the United States and Australia, Tobin found that a small group of mostly male "target" students dominated whole-class activities in both high-achieving and low-achieving classes, and they sometimes dominated small group laboratory work as well (Tobin, 1987). In high-achieving classes, the target students were often boys who called out answers during teacher-led lectures, demonstrations, or discussions, while in low-achieving classes, the target students were sometimes those who interrupted the teacher.

Data from the 2000 survey of science and mathematics education indicate that some teachers view the task of managing all the students in large science classes as a challenge. In 2000, 14 percent of high school science program representatives viewed large classes as a serious problem for science instruction, 20 percent said that student absences were a serious problem, and 5 percent indicated that maintaining discipline was a serious problem (Smith et al., 2002).

In Tobin's observational study, while the teachers focused on responding to or managing the target students, other students rarely asked or answered questions. On the basis of these and other studies, Tobin speculated, "It is entirely likely that high achieving students engage to a greater extent than low achieving students in laboratory activities" (Tobin, 1990, p. 408).

Kelly's (1988) study of gender differences found that male students actively handle laboratory equipment and supplies more frequently than female students. Kahle, Parker, Rennie, and Riley (1993) found that a small group of male students dominated the use of equipment and also sometimes contaminated reagents or otherwise interfered with equipment and materials. More recently, Jovanovic and King (1998) conducted detailed observations of student behavior in six science classrooms with students in grades 5 through 8. The six classrooms were selected on the basis of a competitive process in which exceptional science teachers were nominated by the Fermi National Accelerator Laboratory Education Office in Batavia, Illinois, based on their expertise in hands-on science teaching. The researchers found that boys and girls were equally likely to actively lead small laboratory groups, but that, as members of the groups, boys more often manipulated equipment while girls more often engaged in such passive behaviors as making suggestions or reading directions. Active leadership was a significant predictor of students' attitudes toward science and perception of their abilities in science, regardless of gender. Nevertheless, girls' perceptions of their own science abilities declined over the course of the year as they engaged in passive behaviors more frequently than boys.

The Jovanovic and King study provides evidence in support of Tobin's speculation that high-achieving students engage more actively than others in laboratory activities. The authors found a strong correlation between students' science ability (as measured by a state science assessment) and the frequency of active leadership and equipment manipulation in the laboratory group.

A study of introductory college biology laboratories compared male and female behavior in cookbook laboratory classes with behavior in reformed classes (Russell and French, 2002). In the reformed class, students formed self-selected groups at the beginning of the semester. Before each laboratory period, each student familiarized himself or herself with the activity, developed one or more hypotheses to test, and predicted experimental outcomes. During the laboratory period, the group performed one or more of the planned experiments and wrote a report. The authors of this study conducted detailed observations of students and also surveyed their attitudes and achievement during the semester in which they participated in either the cookbook or the reformed class. They found a positive relationship between time spent manipulating equipment and achievement, as measured by a test of biology content knowledge. They also found that girls participated less frequently in manipulating equipment in both the cookbook and reformed classes, but the gender differences in participation were reduced in the reformed class.

SUMMARY

Research evidence on the current laboratory experiences of U.S. high school students is limited. The few studies available provide information on the amount of time students spend in laboratory activities, the goals of these activities, and how teachers and students act during those activities. Findings from these studies, analyzed in light of information on the educational outcomes of laboratory activities, indicate that access to laboratory experiences is uneven, and the quality of current laboratory experiences is poor for most students.

Most science students in U.S. high schools today participate in typical laboratory experiences. Instead of focusing on clear learning goals, teachers and laboratory manuals often emphasize the procedures to be followed, leaving students uncertain about what they are supposed to learn. Lacking a focus on learning goals related to the subject matter being addressed in the science class, current laboratory experiences often fail to integrate student learning about the processes of science with learning about science content. Few current laboratory experiences incorporate ongoing reflection and discussion between and among the teacher and the students, although there is evidence that such reflection and discussions are essential to help students make meaning out of their laboratory activities. In general, most high school laboratory experiences do not follow the instructional design principles for effectiveness identified by the committee.

Students in schools with higher concentrations of non-Asian minorities spend less time in laboratory instruction than students in other schools, and students in lower level science classes spend less time in laboratory instruction than those enrolled in more advanced science classes. In addition, most high school students participate in a limited range of laboratory activities that do not fully reflect the range of scientists' activities, limiting opportunities for them to gain understanding of the processes of science.

Laboratory experiences for most high school students today appear to have changed little from those observed by Sutman and colleagues in the mid-1990s. Observing many high school and introductory college laboratory experiences, they found (Sutman et al., 1996, pp. 5-6):

> (1) Students experience laboratory based experiences as an add-on to lecture rather than as the "driving force" for later instruction; (2) a very high percentage of the laboratory instructors' time is spent listening to and responding to students' procedural questions, with almost no time available for calling upon strategies designed to develop or strengthen higher order thinking. Post-laboratory experiences almost never include follow-up discussion or analysis of the laboratory findings. At the secondary school level laboratory activities were designed to "fit into" or be completed in a designated period of 45 to 90 minutes. There were never additional opportunities for students to extend the basic study. [R]eports of laboratory experiences were graded and returned to students. The reports were never used diagnostically nor did the grades have significance in determining the final course grades.

In the next chapter, we analyze further evidence about factors contributing to the weakness of current laboratory experiences. That chapter contrasts the types of capacity and support teachers need to lead effective laboratory experiences with the limited capacity and support currently available.

REFERENCES

Atanda, R. (1999). *Do gatekeeper courses expand educational options?* Washington, DC: U.S. Department of Education, National Center for Education Statistics.

Banilower, E.R., Green, S., and Smith, P.S. (2004). *Analysis of data of the 2000 National Survey of Science and Mathematics Education for the Committee on High School Science Laboratories* (September). Chapel Hill, NC: Horizon Research.

College Board. (2001). *AP biology lab manual for teachers.* Princeton, NJ: Author.

College Board. (2005). *Advanced placement report to the nation 2005.* Princeton, NJ: Author. Available at: http://www.apcentral.collegeboard.com/documentlibrary [accessed Feb. 2005].

DeCarlo, C.L., and Rubba, P.A. (1994). What happens during high school chemistry laboratory sessions? A descriptive case study of the behaviors exhibited by three teachers and their students. *Journal of Science Teacher Education, 5*(2), 37-47.

Hammer, D. (1997). Discovery learning and discovery teaching. *Cognition and Instruction, 15*(4), 485-529.

Hegarty-Hazel, E. (Ed.). (1990). *The student laboratory and the science curriculum.* London, England: Rutledge.

Hestenes, D., Jackson, J., Dukerich, L., and Swackhamer, G. (2002). *Modeling instruction in high school physics: Unit 1: Scientific thinking in experimental settings.* Tempe, AZ: Modeling Instruction Program. Available at: http://www.modeling.asu.edu/Modeling-pub/Mechanics_curriculum/1-Sci%20Thinking/ [accessed Feb. 2005].

Hofstein, A., and Lunetta. V. (2004). The laboratory in science education: Foundations for the twenty-first century. *Science Education, 88,* 28-54.

Jovanovic, J., and King, S.S. (1998). Boys and girls in the performance-based science classroom: Who's doing the performing? *American Educational Research Journal, 35*(3), 477-496.

Kahle, J.B., Parker, L.H., Rennie, L.J., and Riley, D. (1993). Gender differences in science education: Building a model. *Educational Psychologist, 28,* 379-404.

Kelly, A. (1988). Sex stereotypes and school science: A three-year follow-up. *Educational Studies, 14,* 151-163.

Lazarowitz, R., and Tamir, P. (1994). Research on using laboratory instruction in science. In D.L. Gabel (Ed.), *Handbook of research on science teaching and learning* (pp. 94-130). New York: Macmillan.

Linn, M.C. (2003). Technology and science education: Starting points, research programs, and trends. *International Journal of Science Education, 25*(6), 727-758.

Linn, M.C., Songer, N.B., and Eylon, B.S. (1996). Shifts and convergences in science learning and instruction. In R. Calfee and D. Berliner (Eds.), *Handbook of educational psychology* (pp. 438-490). Riverside, NJ: Macmillan.

Lunetta, V.N. (1998). The school science laboratory. In B.J. Fraser and K.G. Tobin (Eds.), *International handbook of science education* (pp. 249-262). London, England: Kluwer Academic.

Marx, R.W., Freeman, J.G., Krajcik, J.S., and Blumenfeld, P.C. (1998). Professional development of science teachers. In B.J. Fraser and K.G. Tobin (Eds.), *International handbook of science education* (pp. 667-680). London, England: Kluwer Academic.

Millar, R. (2004). *The role of practical work in the teaching and learning of science:* Paper prepared for the Committee on High School Science Laboratories: Role and Vision, June 3-4, National Research Council, Washington, DC. Available at: http://www7.nationalacademies.org/bose/June3-4_2004_High_School_Labs_Meeting_Agenda.html [accessed April 2005].

National Center for Education Statistics. (2004). *Contexts of elementary and secondary education: Trends in science and mathematics course taking.* Available at: http://www.nces.ed.gov/programs/coe/2004/section4/indicator21.asp [accessed June 2005].

National Research Council. (2002). *Learning and understanding: Improving advanced study of mathematics and science in U.S. high schools.* Washington, DC: National Academy Press. Chemistry Content Panel Report available at: http://www.books.nap.edu/books/NI000403/html/ [accessed Oct., 2004].

National Science Foundation. (2004). *Science and engineering indicators 2004.* Arlington, VA: Author. Available at: http://www.nsf.gov/sbe/srs/seind04/start.htm [accessed Feb. 2005].

Norman, O., Ault, C.R., Bentz, B., and Meskimen, L. (2001). The black-white "achievement gap" as a perennial challenge of urban science education: A sociocultural and historical overview with implications for research and practice. *Journal of Research in Science Teaching, 38*(10), 1101-1114.

Oakes, J. (1990). *Multiplying inequalities: The effects of race, social class, and tracking on opportunities to learn mathematics and science.* Santa Monica, CA: Rand Corporation.

Oakes, J., Gamoran, A., and Page, R. (1992).Curriculum differentiation: Opportunities, outcomes, and meanings. In P. Jackson (Ed.), *Handbook of research on curriculum* (pp. 570- 608). New York: Macmillan.

Oakes, J., Rogers, J., McDonough, P., Solorzano, D., Mehan, H., and Noguera, P. (2000). *Remedying unequal opportunities for successful participation in advanced placement courses in California high schools.* Unpublished paper prepared for the American Civil Liberties Union of Southern California.

Olsen, T.P., Hewson, P.W., and Lyons, L. (1996). Preordained science and student autonomy: The nature of laboratory tasks in physics classrooms. *International Journal of Science Education, 18*(7), 775-790.

O'Sullivan, C.Y., Lauko, M.S., Grigg, W.S., Quian, J., and Zhang, J. (2003). The nation's report card: Science 2000. In *NCES National Assessment of Educational Progress (NAEP).* NCES 2002-452. Washington, DC: U.S. Department of Education.

Perkins, R., Kleiner, B., Roey, S., and Brown, J. (2004). *The High School Transcript Study: A decade of change in curricula and achievement, 1990-2000.* Washington, DC: National Center for Education Statistics. Available at: http://www.nces.ed.gov/pubsearch/pubsinfo.asp?pubid=2004455 [accessed June 2005].

Priestley, W., Priestley, H., and Schmuckler, J. (1997). *The impact of longer term intervention on reforming the approaches to instructions in chemistry by urban teachers of physical and life sciences at the secondary school level.* Paper presented at the National Association for Research in Science Teaching meeting, March 23, Chicago.

Quinn, P. (1996). *NELS: 88 High school seniors' instructional experiences in science and math.* Washington, DC: U.S. Department of Education, National Center for Education Statistics, NCES 95278. Available at: http://www.nces.ed.gov/pubsearch/pubsinfo.asp?pubid=95278 [accessed Dec. 2004].

Russell, C.P., and French, D.P. (2002). Factors affecting participation in traditional and inquiry-based laboratories. *Journal of College Science Teaching, 31,* 225-229.

Shymansky, T.A., and Penick, T.E. (1981). Teacher behavior does make a difference in hands-on science classrooms. *School Science and Mathematics, 81,* 412-423.

Smith, P.S., Banilower, E.R., McMahon, K.C., and Weiss, I.R. (2002). *The National Survey of Science and Mathematics Education: Trends from 1977 to 2000.* Chapel Hill, NC: Horizon Research. Available at: http://www.horizon-research.com/reports/2002/2000survey/trends.php [accessed May 2005].

Sutman, F.X., Schmuckler, J.S., Hilosky, A.B., Priestley, H.D., and Priestley, W.J. (1996). *Seeking more effective outcomes from science laboratory experiences (grades 7-14): Six companion studies.* Summary of multiple paper presentation at the annual meeting of the National Association for Research in Science Teaching, April 1, St. Louis, MO.

Tobin, K. (1987). Forces which shape the implemented curriculum in high school science and mathematics. *Teaching and Teacher Education, 3*(4), 287-298.

Tobin, K. (1990). Research on science laboratory activities: In pursuit of better questions and answers to improve learning. *School Science and Mathematics, 90*(5), 403-418.

Tobin, K., and Gallagher, J. (1987). What happens in high school science classrooms? *Journal of Curriculum Studies, 19*(6), 549-560.

Weiss, I., Banilower, E., McMahon, K., and Smith, P.S. (2001). *Report of the 2000 Survey of Mathematics and Science Education.* Chapel Hill, NC: Horizon Research. Available at: http://www.2000survey.horizon-research.com/reports/status.php [accessed Dec. 2004].

5

Teacher and School Readiness for Laboratory Experiences

Key Points

- *Leading laboratory experiences is a demanding task requiring teachers to have sophisticated knowledge of science content and process, how students learn science, assessment of students' learning, and how to design instruction to support the multiple goals of science education.*
- *Pre-service education and in-service professional development for science teachers rarely address laboratory experiences and do not provide teachers with the knowledge and skills needed to lead laboratory experiences.*
- *There are promising examples of teacher professional development focused on laboratory experiences. Further research is needed to inform design of professional development that can effectively support improvements in teachers' laboratory instruction.*
- *School administrators play a critical role in supporting the successful integration of laboratory experiences in high school science by providing improved approaches to professional development and adequate time for teacher planning and implementation of laboratory experiences.*

This chapter describes some of the factors contributing to the weakness of current laboratory experiences. We begin by identifying some of the knowledge and skills required to lead laboratory experiences aligned with the goals and design principles we have identified. We then compare the desired skills and knowledge with information about the current skills and knowledge of high school science teachers. We then go on to describe approaches to supporting teachers and improving their capacity to lead laboratory experiences through improvements in professional development and use of time. The final section concludes that there are many barriers to improving laboratory teaching and learning in the current school environment.

TEACHERS' CAPACITY TO LEAD LABORATORY EXPERIENCES

In this section, we describe the types of teacher knowledge and skills that may be required to lead a range of laboratory experiences aligned with our design principles, comparing the required skills with evidence about the current state of teachers' knowledge and skills. We then present promising examples of approaches to enhancing teachers' capacity to lead laboratory experiences.

Teacher Knowledge for a Range of Laboratory Experiences

Teachers do not have sole responsibility for carrying out laboratory experiences that are designed with clear learning outcomes in mind, thoughtfully sequenced into the flow of classroom science instruction, integrating the learning of science content and process, and incorporating ongoing student reflection and discussion, as suggested by the research. Science teachers' behavior in the classroom is influenced by the science curriculum, educational standards, and other factors, such as time constraints and the availability of facilities and supplies. Among these factors, curriculum has a strong influence on teaching strategies (Weiss, Pasley, Smith, Banilower, and Heck, 2003). As discussed in Chapters 2 and 3, there are curricula that integrate laboratory experiences into the stream of instruction and follow the other instructional design principles. To date, however, few high schools have adopted such research-based science curricula, and many teachers and school administrators are unaware of them (Tushnet et al., 2000; Baumgartner, 2004).

Studies of the few schools and teachers that have implemented research-based science curricula with embedded laboratory experiences have found that engaging teachers in developing and refining the curricula and in professional development aligned with the curricula leads to increases in students' progress toward the goals of laboratory experiences (Slotta, 2004). These studies confirm earlier research findings that even the best science curriculum cannot "teach itself" and that the teacher's role is central in helping students build understanding from laboratory experiences and other science learning activities (Driver, 1995).

Playing this critical role requires that teachers know much more than how to set up equipment, carry out procedures, and manage students' physical activities. Teachers must consider how to select curriculum that integrates laboratory experiences into the stream of instruction and how to select individual laboratory activities that will fit most appropriately into their science classes. They must consider how to clearly communicate the learning goals of the laboratory experience to their students. They must address the challenge of helping students to simultaneously develop scientific reasoning, master science subject matter and progress toward the other goals of laboratory experiences. They must guide and focus ongoing discussion and reflection with individuals, laboratory groups, and the entire class. At the same time, teachers must address logistical and practical concerns, such as obtaining and storing supplies and maintaining laboratory safety.

Teachers require several types of knowledge to succeed in these multiple activities, including (1) science content knowledge, (2) pedagogical content knowledge, (3) general pedagogical knowledge, and (4) knowledge of appropriate assessment techniques to measure student learning in laboratory education.

Science Content Knowledge

Helping students attain the learning goals of laboratory experiences requires their teachers to have broad and deep understanding of both the processes and outcomes of scientific research. The degree to which teachers themselves have attained the goals we speak of in this report is likely to influence their laboratory teaching and the extent to which their students progress toward these goals.

Teachers require deep conceptual knowledge of a science discipline not only to lead laboratory experiences that are designed according to the research, but also to lead a full range of laboratory experiences reflecting the range of activities of scientists (see Chapter 1). Deep disciplinary expertise is necessary to help students learn to use laboratory tools and procedures and to make observations and gather data. It is necessary even to lead students in activities designed to verify existing scientific knowledge. Case studies of laboratory teaching show that laboratory activities designed to verify known scientific concepts or laws may not always go forward as planned (Olsen et al., 1996). Guiding students through the complexity and ambiguity of empirical work—including verification work—requires deep knowledge of the specific science concepts and science processes involved in such work (Millar, 2004).

As teachers move beyond laboratory experiences focusing on tools, procedures, and observations to those that engage students in posing a research question or in building and revising models to explain their observations, they require still deeper levels of science content knowledge (Windschitl, 2004; Catley, 2004). When students have more freedom to pose questions or to identify and carry out procedures, they require greater guidance to ensure that their laboratory activities help them to master science subject matter and progress toward the other goals of laboratory experiences. Teachers require a deep understanding of scientific processes in order to guide students' procedures and formulation of research questions, as well as deep understanding of science concepts in order to guide them toward subject matter understanding and other learning goals. Engaging students in analysis of data gathered in the laboratory and in developing and revising explanatory models for those data requires teachers to be familiar with students' practical equipment skills and science content knowledge and be able to engage in sophisticated scientific reasoning themselves.

Pedagogical Content Knowledge

To lead laboratory experiences that incorporate ongoing student discussion and reflection and that focus on clear, attainable learning goals, teachers require pedagogical content knowledge. This is knowledge drawn from learning theory and research that helps to explain how students develop understanding of scientific ideas. Pedagogical content knowledge may include knowing what theories of natural phenomena students may hold and how their ideas may differ from scientific explanations, knowledge of the ideas appropriate for children to explore at different ages, and knowledge of ideas that are prerequisites for their understanding of target concepts. Shulman (1986, p. 8) has defined pedagogical content knowledge as:

> [A] special amalgam of content and pedagogy that is uniquely the province of teachers, their own form of professional understanding. . . . [I]t represents the blending of content and pedagogy into an understanding of how particular topics, problems, or issues are organized, represented and adapted to the diverse interests and abilities of learners, and presented for instruction.

Deng (2001) describes pedagogical content knowledge for science teachers as an understanding of key scientific concepts that is somewhat different from that of a scientist. He suggests that a high school physics teacher should know concepts or principles to emphasize when introducing high school students to a particular topic (p. 264). For example, the teacher might use descriptive or qualitative language or images to convey concepts related to light, such as reflection, transmission, and absorption. In contrast, a physicist might use mathematics to describe or represent the reflection, transmission, and absorption of light.

Pedagogical content knowledge can help teachers and curriculum developers identify attainable science learning goals, an essential step toward designing laboratory experiences with clear learning goals in mind. For example, in developing the Computers as Learning Partners science curriculum unit, Linn and colleagues researched how well models of thermodynamics at various levels of abstraction supported students' learning. They found that a heat-flow model was better able to connect to middle school students' knowledge about heat and temperature than a molecular-kinetic model (Linn, Davis, and Bell, 2004). Linn describes aspects of the model as pragmatic principles of heat that are "more accessible goals than the microscopic view of heat that is commonly taught" (Linn, 1997, p. 410). The research team focused the curriculum on helping students understand these principles, including flow principles, rate principles, total heat flow principles, and an integration principle.

The importance of pedagogical content knowledge challenges assumptions about what science teachers should know in order to help students attain the goals of laboratory experiences. Specifically, it challenges the assumption that having a college degree in science, by itself, is sufficient to teach high school science. Familiarity with the evidence or principles of a complex theory does not ensure that a teacher has a sound understanding of concepts that are meaningful to high school students and that she or he will be capable of leading students to change their ideas by critiquing each others' investigations as they make sense of phenomena in their everyday lives. Expertise in science alone also does not ensure that teachers will be able to anticipate which concepts will pose the greatest difficulty for students and design instruction accordingly.

General Pedagogical Knowledge

In addition to science content knowledge and pedagogical content knowledge, teachers also need general pedagogical knowledge in order to moderate ongoing discussion and reflection on laboratory activities, and supervise group work.

Knowledge of children's mental and emotional development, of teaching methods, and how best to communicate with children of different ages is essential for teachers to help students build meaning based on their laboratory experiences.

Because many current science teachers have demographic backgrounds different from their students (Lee, 2002; Lynch, Kuipers, Pyke, and Szeze, in press), the ability to communicate across barriers of language and culture is an increasingly important aspect of their general pedagogical knowledge. Lee and Fradd (1998) and others observe that some scientific values and attitudes are found in most cultures (e.g., wonder, interest, diligence, persistence, imagination, respect toward nature); others are more characteristic of Western science. For example, Western science promotes a "critical and questioning stance," and "these values and attitudes may be discontinuous with the norms of cultures that favor cooperation, social and emotional support, consensus building, and acceptance of the authority" (p. 470). Knowledge of students' cultures and languages and the ability to communicate across cultures are necessary to carry out laboratory experiences that build on diverse students' sense of wonder and engage them in science learning.

Knowledge of Assessment

Focusing laboratory experiences on clear learning goals requires that teachers understand assessment methods so they can measure and guide their students' progress toward those goals. To be successful in leading students across the range of laboratory experiences we have described, teachers must choose laboratory experiences that are appropriate at any given time. To make these choices, they must be aware not only of their own capabilities, but also of students' needs and readiness to engage in the various types of laboratory experiences.

Teacher awareness of students' science needs and capabilities may be enhanced through ongoing formative assessment. Formative assessment, that is, continually assessing student progress in order to guide further instruction, appears to enhance student attainment of the goals of laboratory education. Teachers need to use data drawn from conversations, observations, and previous student work to make informed decisions about how to help them move toward desired goals. This is not a simple task (National Research Council, 2001b, p. 79):

> To accurately gauge student understanding requires that teachers engage in questioning and listen carefully to student responses. It means focusing the students' own questions. It means figuring out what students comprehend by listening to them during their discussions about science. They need to carefully consider written work and what they observe while students engage in projects and investigations. The teacher strives to fathom what the student is saying and what is implied about the student's knowledge in his or her statements, questions, work and actions. Teachers need to listen in a way that goes well beyond an immediate right or wrong judgment.

Methods of assessing student learning in laboratory activities include systematically observing and evaluating students' performance in specific laboratory tasks and longer term laboratory investigations. Teachers also need to know how to judge the quality of students' oral presentations, laboratory notebooks, essays, and portfolios (Hein and Price, 1994; Gitomer and Duschl, 1998; Harlen, 2000, 2001). To lead effective laboratory experiences, science teachers should know how to use data from all of these assessment methods in order to reflect on student progress and make informed decisions about which laboratory activities and teaching approaches to change, retain, or discard (National Research Council, 2001b; Volkman and Abell, 2003).

Teachers' Knowledge in Action

Teachers draw on all of the types of knowledge listed above—content knowledge, pedagogical content knowledge, general pedagogical knowledge, and knowledge of assessment—in their daily work of planning and leading instruction. Formulating research questions appropriate for a science classroom and leading student discussions are two important places where the interaction of the four types of knowledge is most evident.

In developing an investigation for students to pursue, teachers must consider their current level of knowledge and skills, the range of possible laboratory experiences available, and how a given experience will advance their learning. Teachers need to decide what kind of phenomena are important and appropriate for students to study as well as the degree of structure their students require.

Currently, teachers rarely provide opportunities for students to participate in formulating questions to be addressed in the laboratory. Perhaps this is because, among scientists, decisions about the kinds of questions to be asked and the kinds of answers to be sought are often developed by the scientific community rather than by an isolated individual (Millar, 2004). Only a few high school students are sufficiently advanced in their knowledge of science to serve as an effective "scientific community" in formulating such questions. Guiding students to formulate their own research questions and design appropriate investigations requires sophisticated knowledge in all four of the domains we have identified.

The teacher's ability to use sophisticated questioning techniques to bring about productive student-student and student-teacher discussions in all phases of the laboratory activity is a key factor in the extent to which the activity attains its goals (Minstrell and Van Zee, 2003). However, formulating such questions can be difficult (National Research Council, 2001a, 2001b). To succeed at it and ask the types of higher level and cognitively based questions that appear to support student learning, teachers must have considerable science content knowledge and science teaching experience (McDiarmid, Ball, and Anderson, 1989; Chaney, 1995; Sanders and Rivers, 1996; Hammer, 1997).

The teacher's skills in posing questions and leading discussions affect students' ability to build meaning from their laboratory experiences. As students analyze observations from the laboratory in search of patterns or explanations, develop and revise conjectures, and build lines of reasoning about why their proposed claims or explanations are or are not true, the teacher supports their learning by conducting sense-making discussions (Mortimer and Scott, 2003; van Zee and Minstrell, 1997; Hammer, 1997; Windschitl, 2004; Bell, 2004; Brown and Campione, 1998; Bruner, 1996; Linn, 1995; Lunetta, 1998; Clark, Clough, and Berg, 2000; Millar and Driver, 1987). In these discussions, the teacher helps students to resolve dissonances between the way they initially understood a phenomenon and the new evidence. But those connections are not enough: science sense-making discourse must also help students to develop understanding of a given science concept and create links between theory and observable phenomena. The teachers' skills in posing questions and leading discussions also help students to effectively and accurately communicate their laboratory activities and the science sense they make from them, using appropriate language, scientific knowledge, mathematics, and other intellectual modes of communication associated with a particular science discipline.

Currently, few teachers lead this type of sense-making discussion (Smith, Banilower, McMahon, and Weiss, 2002). This lack of discussion may be due to the fact that high school science teachers depend heavily on the use of textbooks and accompanying laboratory manuals (Smith et al., 2002), which rarely include discussions. It may also be because teachers lack the content knowledge, pedagogical content knowledge, general pedagogical knowledge, and knowledge of assessment required to lead such discussions (Maienschein, 2004; Windschitl, 2004). Supporting classroom discussions may be particularly challenging for teachers who work with a very diverse student population in a single classroom, or those who have a different cultural background from their students (see Tobin, 2004).

Current State of Teacher Knowledge: Preservice Education

The available evidence indicates that the current science teaching workforce lacks the knowledge and skills required to lead a range of effective laboratory experiences.

Uneven Qualifications of Science Teachers

A series of studies conducted over the past several decades has shown that teachers are one of the most important factors influencing students' educational outcomes (Ferguson, 1998; Goldhaber, 2002; Goldhaber, Brewer, and Anderson, 1999; Hanushek, Kain, and Rivkin, 1999; Wright, Horn, and Sanders, 1997). However, experts do not agree on which aspects of teacher quality—such as having an academic major in the subject taught, holding a state teaching certificate, having a certain number of years of teaching experience, or other unknown factors—contribute to their students' academic achievement (Darling-Hammond, Berry, and Thoreson, 2001; Goldhaber and Brewer, 2001). Generally, the body of research is weak, and the effects of teacher quality on student outcomes are small and specific to certain contexts.

Studies focusing specifically on science teacher quality and student achievement are somewhat more conclusive. Researchers generally agree that the teachers' academic preparation in science has a positive influence on students' science achievement (U.S. Department of Education, 2000; National Research Council, 2001a). One study found that having an advanced degree in science was associated with increased student science learning from the 8th to the 10th grade (Goldhaber and Brewer, 1997). The National Research Council (NRC) Committee on Science and Mathematics Teacher preparation stated that "studies conducted over the past quarter century increasingly point to a strong correlation between student achievement in K-12 science and mathematics and the teaching quality and level of knowledge of K-12 teachers of science and mathematics" (National Research Council, 2001a, p. 4).

A teacher's academic science preparation appears to affect student science achievement generally. Strong academic preparation is also essential in helping teachers develop the deep knowledge of science content and science processes needed to lead effective laboratory experiences. However, many high school teachers currently lack strong academic preparation in a science discipline. Data from the National Center for Education Statistics (2004) show variation in teacher qualifications from one science discipline to another. In 1999-2000, 39.4 percent of all physics teachers in public high schools had neither a major nor a minor in physics, 59.9 percent of all public high school geology teachers lacked a major or minor in geology, 35.7 percent of chemistry teachers lacked a major or minor in that field, and 21.7 percent of biology teachers had neither a major nor a minor in biology (National Center for Education Statistics, 2004). Another analysis of the data from the National Center for Education Statistics found that students in high schools with higher concentrations of minority students and poor students were more likely than students in other high schools to be taught science by a teacher without a major or minor in the subject being taught (U.S. Department of Education, 2004).

The inequities in the availability of academically prepared teachers may pose a serious challenge to minority and poor students' progress toward the goals of laboratory experiences. Teachers lacking a science major may be less likely to engage students in any type of laboratory experience and may be less likely to provide more advanced laboratory experiences, such as those that engage the students in posing research questions, in formulating and revising scientific models, and in making scientific arguments. These limits, in turn, could contribute to lower science achievement, especially among poor and minority students.

Uneven Quality of Preservice Science Education

Even teachers who have majored in science may be limited in their ability to lead effective laboratory experiences, because their undergraduate science preparation provided only weak knowledge of science content and included only weak laboratory experiences. Research conducted in teacher education programs provides some evidence of the quality of preservice science education (Windschitl, 2004). One theme that emerges from such research is that the content knowledge gained from undergraduate work is often superficial and not well integrated. The traditional didactic pedagogy to which teacher candidates are exposed in university science courses equips learners with only minimal conceptual understandings of their science disciplines (Duschl, 1983; Gallagher, 1991; Pomeroy, 1993, cited in Windschitl, 2004). Many preservice teachers hold serious misconceptions about science that are similar to those held by their students (Anderson, Sheldon, and Dubay, 1990; Sanders, 1993; Songer and Mintzes, 1994; Westbrook and Marek, 1992, all cited in Windschitl, 2004).

The limited evidence available indicates that some undergraduate science programs do not help future teachers develop full mastery of science subject matter. In a year-long study of prospective biology teachers (Gess-Newsome and Lederman, 1993), the participants reported never having thought about the central ideas of biology or the interrelationships among the topics. The teachers, all biology majors, could only list the courses they had taken as a way to organize their fields. They appeared to have little understanding of the field writ large. They knew little about how various ideas were related to each other, nor could they readily explain the overall content and character of biology. Over the course of a year's worth of pedagogical preparation and field experiences, the new teachers began to reorganize their knowledge of biology according to how they thought it should be taught. These findings confirm those from a substantial literature on arts and sciences teaching in colleges and universities, which has clearly documented that both elementary and secondary teachers lack a deep and connected conceptual understanding of the subject matter they are expected to teach (Kennedy, Ball, McDiarmid, and Schmidt, 1991; McDiarmid, 1994).

Undergraduate science students, including preservice teachers, engage in a limited range of laboratory experiences that do not follow the principles of instructional design identified in Chapter 3. The research described above indicates that undergraduate laboratory experiences do not integrate learning of science content and science processes in ways that lead to deep conceptual understanding of science subject matter. Other studies report that undergraduate laboratory work consists primarily of verification activities, with few opportunities for ongoing discussion and reflection on how scientists evaluate new knowledge (e.g., Trumbull and Kerr, 1993, cited in Windschitl, 2004). The research also indicates that undergraduate laboratory work, like the laboratory experiences of high school students, often focuses on detailed procedures rather than clear learning goals (Hegarty-Hazel, 1990; Sutman, Schmuckler, Hilosky, Priestley, and Priestley, 1996).

One study illustrates undergraduate students' lack of exposure to the full range of scientists' activities, and the potential benefits of engaging them in a broader range of experiences. A professor engaged upper level chemistry majors in trying to create a foolproof laboratory activity to illustrate the chemistry of amines for introductory students. Students were asked to survey the literature for methods to reduce aromatic nitro compounds to the corresponding amines. They found a large number of preparations, tried each one out, and identified one method as most likely to succeed with the introductory students. However, the students were surprised that methods taken from the literature did not always work. Their previous, closely prescribed laboratory experiences had not helped them to understand that there are many different ways to effect a particular chemical transformation. More than 90 percent of the class indicated that the experiment was highly effective in demonstrating the difficulty of scientific investigations and the possibility of failure in science (Glagovich and Swierczynski, 2004).

Similarly, Hilosky, Sutman, and Schmuckler (1998) observe that prospective science teachers' laboratory experiences provide procedural knowledge but few opportunities to integrate science investigations with learning about the context of scientific models and theories. In a study of 100 preservice science teachers, only 20 percent reported having laboratory experiences that gave them opportunities to ask their own questions and to design their own science investigations (Windschitl, 2004). A study of a much smaller sample of teachers yielded similar findings (Catley, 2004).

It appears that the uneven quality of current high school laboratory experiences is due in part to the preparation of science teachers to lead these experiences. Science teachers may be modeling instructional practices they themselves witnessed or experienced firsthand as students in college science classes. Clearly, their preservice experiences do not provide the skills and knowledge needed to select and effectively carry out laboratory experiences that are appropriate for reaching specific science learning goals for a given group of students.

Professional Development for Laboratory Teaching

Current professional development for science teachers is uneven in quantity and quality and places little emphasis on laboratory teaching. Requirements for professional development of in-service science teachers differ widely from state to state. Most states do not regulate the quality and content of professional development required for renewal of teaching certificates (Hirsch, Koppich, and Knapp, 2001). Typically, states require only that teachers obtain post-baccalaureate credits within a certain period of time after being hired and then earn additional credits every few years thereafter.

Few professional development programs for science teachers emphasize laboratory instruction. In reviewing the state of biology education in 1990, an NRC committee concluded that few teachers had the knowledge or skill to lead effective laboratory experiences and recommended that "major new programs should be developed for providing in-service education on laboratory activities" (National Research Council, 1990, p. 34). However, a review of the literature five years later revealed no widespread efforts to improve laboratory education for either preservice or in-service teachers (McComas and Colburn, 1995). The authors of the review found that, when laboratory education is available, it focuses primarily on the care and use of laboratory equipment and laboratory safety. In addition, they found that commercially available laboratory manuals failed to provide "cognitively challenging activities" that might help to bridge the gap between teachers' lack of knowledge and improved laboratory experiences (McComas and Colburn, 1995, p. 120).

The limited quality and availability of professional development focusing on laboratory teaching is a reflection of the weaknesses in the larger system of professional development for science teachers. Data from a 2000 survey of science and mathematics education indicate that most current science teachers participate infrequently in professional development activities, and that many teachers view these activities as ineffective (Hudson, McMahon, and Overstreet, 2002). For example, among high school teachers who had participated in professional development aimed at learning to use inquiry-oriented teaching strategies, 25 percent indicated that this professional development had little or no impact, and 48 percent reported that the professional development merely confirmed what they were already doing. Other studies have also found that most teachers do not experience sustained professional development and that they view it as ineffective (Windschitl, 2004). In many cases "teachers ranked in-service training as their *least effective* source of learning" (Windschitl, 2004, p. 16; emphasis in original).

Potential of Professional Development for Improved Laboratory Teaching

Despite the weakness of current professional development for laboratory teaching, a growing body of research indicates that it is possible to develop and implement professional development that would support improved laboratory teaching and learning. Most current professional development for science teachers, such as the activities that had little impact on the teaching strategies among teachers responding to the 2000 survey, is ad hoc. It often consists mostly of one-day (or shorter) workshops focusing on how-to activities that are unlikely to challenge teachers' beliefs about teaching and learning that support their current practice (DeSimone, Garet, Birman, Porter, and Yoon, 2003).

In contrast to these short, ineffective approaches, consensus is growing in the research about key features of high-quality professional development for mathematics and science teachers (DeSimone, Porter, Garet, Yoon, and Birman, 2002; DeSimone et al., 2003, p. 10):

- New forms of professional development (i.e., study group, teacher network, mentoring, or task force, internship, or individual research project with a scientist) in contrast to the traditional workshop or conference.
- Duration (total contact hours, span of time).
- Participation of groups of teachers from the same school, department, or grade.
- A focus on deepening teachers' knowledge of science or mathematics.
- Active learning opportunities focused on analysis of teaching and learning.
- Coherence (consistency with teachers' goals, state standards, and assessments).

Loucks-Horsley, Love, Stiles, Mundry, and Hewson (2003) provide a detailed design framework for professional development and descriptions of case studies, identifying strategies for improving science teaching that may be applicable to improving laboratory teaching.

DeSimone and others conducted a three-year longitudinal study of professional development in science and mathematics provided by school districts. They surveyed a sample of 207 teachers in 30 schools, 10 districts, and 5 states to examine features of professional development and its effects on teaching practice from 1996 to 1999 (DeSimone et al., 2002). The study examined the relationship between professional development and teaching practice in terms of three specific instructional practices: (1) the use of technology, (2) the use of higher order instructional methods, and (3) the use of alternative assessment. The investigators found that professional development focused on specific instructional practices increased teachers' use of these practices in the classroom. Results of the study also confirmed the effectiveness of providing active learning opportunities.

Other studies indicate that high-quality professional development can encourage and support science teachers in leading a full range of laboratory experiences that allow students to participate actively in formulating research questions and in designing and carrying out investigations (Windschitl, 2004). Research on teachers using a science curriculum that integrates laboratory experiences into the stream of instruction indicates that repeated practice with such a curriculum, as well as time for collaboration and reflection with professional colleagues, leads teachers to shift from focusing on laboratory procedures to focusing on science learning goals (Williams, Linn, Ammon, and Gearheart, 2004). One study indicated that significant change in teaching practice required about 80 hours of professional development (Supovitz and Turner, 2000). Teachers who had engaged in even more intensive professional development, lasting at least 160 hours, were most likely to employ several teaching strategies aligned with the design principles for effective laboratory experiences identified in the research. These strategies included arranging seating to facilitate student discussion, requiring students to supply evidence to support their claims, encouraging students to explain concepts to one another, and having students work in cooperative groups.

A study of Ohio's Statewide Systemic Initiative in science and mathematics also confirmed that sustained professional development, over many hours, is required to change laboratory teaching practices (Supovitz, Mayer, and Kahle, 2000, cited in Windschitl, 2004, p. 20): "A highly intensive (160 hours) inquiry-based professional development effort changed teachers' attitudes towards reform, their preparation to use reform-based practices, and their use of inquiry-based teaching practices. . . . These changes persisted several years after the teachers concluded their professional development experiences."

Examples of Professional Development Focused on Laboratory Teaching

The committee identified a limited portfolio of examples of promising approaches to professional development that may support teachers in leading laboratory experiences designed with clear learning outcomes in mind, thoughtfully sequenced into the flow of classroom science instruction, integrating the learning of science content and process, and incorporating ongoing student reflection and discussion. School districts, teachers, and others may want to consider these examples, but further research is needed to determine their scope and effectiveness.

Laboratory Learning: An Inservice Institute. McComas and Colburn (1995) established an inservice program called Laboratory Learning: An Inservice Institute, which incorporated some of the design elements that support student learning in laboratory experiences. The contents of the institute were developed on the basis of in-depth field interviews and literature reviews to tap the practical knowledge of experienced science teachers. This body of knowledge addressed the kind of laboratory instruction given to students, consideration of students with special needs, supportive teaching behaviors, models to engage students working in small groups, the sequencing of instruction, and modes of assessment (p. 121). Teacher participants at the institute experienced firsthand learning as students in several laboratory sessions led by high school instructors who were regarded as master laboratory teachers. The institute included a blend of modeling, small group work, cooperative learning activities, and theoretical and research-based suggestions (p. 122).

This professional development institute also incorporated ongoing opportunities for discussion and reflection. It was implemented over four day-long Saturday sessions spread over a semester. Between sessions, teacher participants reflected on what they were learning and applied some of it in their classrooms, following the active learning approach suggested by the research on professional development for science teachers. The teachers participated in and analyzed practical laboratory activities, studied theoretical underpinnings of the science education they were receiving, and learned about safety issues during hands-on activity. Reporting on a post-institute survey, McComas and Colburn note that "a surprising number of teachers felt that the safety sessions were most important" (p. 121) (no numbers were reported). Institute participants also asked for more discussion of assessment methods for laboratory teaching, including the role of video testing, and also recommended inclusion of sessions that address teaching science laboratory classes on a small budget.

13-Week Science Methodology Course. A science methodology course for middle and high school teachers offered experience in using the findings from laboratory investigations as the driving force for further instruction (Priestley, Priestly, and Schmuckler, 1997). The design of this professional development program incorporated the principle of integrating laboratory experiences into the stream of instruction and the goal of providing a full range of laboratory experiences, including opportunities for students to participate in developing research questions and procedures.

In this program, faculty modeled "lower-level inquiry-oriented instruction" focused on short laboratory sessions with limited lecturing and no definitions of terms. They also modeled longer postlaboratory activities focused on using student data and observations as the engine for further instruction. In doing so, they showed teachers how laboratory experiences can be sequenced into a flow of science instruction in order to integrate student learning of science content and science processes. After completion of the course, teachers' classroom behaviors were videotaped and analyzed against "traditional" and "reformed" instructional strategies. Participant teachers were also interviewed. The authors concluded that professional development activities that are short-term interventions have virtually no effect on teachers' behaviors in leading laboratory experiences. They also concluded that longer term interventions—13 weeks in this case—result in some change in the instructional strategies teachers use.

Project ICAN: Inquiry, Context, and Nature of Science. Project ICAN includes an intensive three-day summer orientation for science teachers followed by full-day monthly workshops from September through June, focusing on the nature of science and scientific inquiry. The program was designed in part to address weakness in science teachers' understanding of the nature of science, which was documented in earlier research (Khalic and Lederman, 2000; Schwartz and Lederman, 2002). This earlier research indicated that, just as engaging students in laboratory experiences in isolation led to little or no increase in their understanding of the nature of science, engaging prospective or current science teachers in laboratory activities led to little or no increase in their understanding of the nature of science. Professional development and preservice programs that combined laboratory experiences with instruction about the key concepts of the nature of science and engaged teachers in reflecting on their experiences in light of those concepts were more successful in developing improved understanding (Khalic and Lederman, 2000).

In the ICAN program, teachers participate in science internships with working scientists as one element in a larger program of instruction that includes an initial orientation and monthly workshops. These workshops include microteaching (peer presentation) sessions. Program faculty report that many teachers tend to dwell on hands-on activities with their students at the expense of linking them with the nature of science and with abilities associated with scientific inquiry. They further report (Lederman, 2004, p. 8):

> By observing practicing scientists and writing up their reflections, teachers gained insight into what scientists do in various research areas, such as crystallization, vascular tissue engineering, thermal processing of materials, nutrition, biochemistry, molecular biology, microbiology, protein purification and genetics. . . . Periodic checks indicated that the science internship helped teachers improve their understanding of [the nature of science] and [science inquiry]. For example, teachers realized that there is no unique method called "the scientific method," after comparing the methods used in different labs, such as a biochemistry lab, engineering lab, and zoos. It was also clear that teachers' enhanced their understanding of science subject matter specific to the lab they experienced.

The Biological Sciences Curriculum Study. The Biological Sciences Curriculum Study, a science curriculum development organization, has long been engaged in the preservice education of science teachers and also offers professional development for inservice teachers. The group employs a variety of long-term strategies, such as engaging teachers in curriculum development and adaptation, action research, and providing on-site support by lead teachers (Linn, 1997; Lederman, 2004). Research on the efficacy of strategies used for professional development related specifically to laboratory experiences, however, is not readily available.

Professional Development Partnerships with the Scientific Community. Scientific laboratories, college and university science departments, and science museums have launched efforts to support high school science teachers in improving laboratory teaching. For example, the U.S. Department of Energy (DOE) launched its Laboratory Science Teacher Professional Development Program in 2004. Building on existing teacher internship programs at several of the national laboratories, the program will engage teachers as summer research associates at the laboratories, beginning with a four-week stint the first summer, followed by shorter two-week internships the following two summers (U.S. Department of Energy, 2004). Qualified high school teachers will have opportunities to work and learn at the Argonne, Brookhaven, Lawrence Berkeley, Oak Ridge, and Pacific Northwest National Laboratories and at the National Renewable Energy Laboratory. The Fermi National Accelerator Laboratory has provided professional development programs for science teachers for several years (Javonovic and King, 1998).

With the support of the Howard Hughes Medical Institute (HHMI), several medical colleges and research institutions provide laboratory-based science experiences for science teachers and their students. For example, HHMI has funded summer teacher training workshops at the Cold Spring Harbor Laboratory for many years, and also supports an ongoing partnership between the Fred Hutchinson Cancer Research Center and the Seattle, Washington, public schools (Fred Hutchinson Cancer Research Center, 2003). In the Seattle program, teachers attend a 13-day summer workshop in which they work closely with each other, master teachers, and program staff to develop expertise in molecular biology. They also spend a week doing laboratory research with a scientist mentor at the Fred Hutchinson Center or one of several other participating public and private research institutions in Seattle. During the school year, teachers may access kits of materials supporting laboratory experiences that use biomedical research tools.

Summer research experiences that may enhance science teachers' laboratory teaching need not take place in a laboratory facility. At Vanderbilt University, Catley conducts a summer-long course on research in organismal biology. Teachers design and carry out an open-ended field research project of their own choosing. Catley (2004) reports that "having gone through the process of frustration, false starts and the elation of completion, [the teachers] came away with a deeper understanding of how inquiry works and a sense of empowerment. They felt confident to guide their students through the same process, where there is no 'right answer.'"

It is unclear whether these and other ad hoc efforts to provide summer research experiences reach the majority of high school science teachers. Although no national information is available about high school teachers' participation in laboratory internship programs, a recent survey found that only 1 in 10 novice elementary school teachers had participated in internship programs in which they worked directly with scientists or engineers. Among those who had, an overwhelming majority said the experience had helped them better understand science content and improved both their teaching practice and their enthusiasm (Bayer Corporation, 2004). Further research is needed to assess the extent to which such programs help teachers develop the knowledge and skills required to lead laboratory experiences in ways that help students master science subject matter and progress toward other science learning goals.

Providing Expert Assistance to Schools and Teachers. In addition to the many programs to increase teachers' knowledge and abilities discussed above, the scientific community sometimes engages scientists to work directly with students. For example, Northeastern University has established a program called RE-SEED (Retirees Enhancing Science Education through Experiments and Demonstration), which arranges for engineers, scientists, and other individuals with science backgrounds to assist middle school teachers with leading students in laboratory experiences. Volunteers receive training, a sourcebook of activities appropriate for middle school students, a kit of science materials, and a set of videotapes. To date, over 400 RE-SEED volunteers have worked with schools in 10 states. A survey of students, teachers, and volunteers yielded positive results. Large majorities of students indicated that the program had increased their interest in science, while large majorities of teachers said they would recommend the program to other teachers and that the volunteers had had a beneficial effect on their science teaching. Among the volunteers, 97 percent said they would recommend RE-SEED to a colleague, and most said that the training, placement in schools, and support from staff had made their time well spent (Zahopoulos, 2003).

The California Institute of Technology has a program to help scientists and graduate students work with teachers in elementary school classrooms in the Pasadena school district. The Chemistry Department of City College (City University of New York) places undergraduate science and engineering majors in middle school classrooms to assist teachers during laboratory activities and learn classroom management from the teachers. Once again, little information is available on the effectiveness of these efforts. Further research is needed to evaluate these and other efforts to link scientists with K-12 education.

We do not yet know how best to develop the knowledge and skills that teachers require to lead laboratory experiences that help students master science subject matter, develop scientific reasoning skills, and attain the other goals of laboratory education. Further research is needed to examine the scope and effectiveness of the many individual programs and initiatives. Because efforts to improve teachers' ability to lead improved laboratory experiences are strongly influenced by the organization and administration of their schools, the following section addresses this larger context.

SUPPORTING LABORATORY TEACHING

The poor quality of laboratory experiences of most high school students today results partly from the challenges that laboratory teaching and learning pose to school administrators. In this section we describe the difficulty school administrators encounter when they try to support effective laboratory teaching.

Supporting Teachers with Professional Development

School administrators have a strong influence on whether high school science teachers receive the professional development opportunities needed to develop the knowledge and skills we have identified. Providing more focused, effective, and sustained professional development activities for more science teachers requires not only substantial financial resources and knowledge of effective professional development approaches, but also a coherent, coordinated approach at the school and district level.

Some school and school district officials may be reluctant to invest in sustained professional development for science teachers because they fear losing their investments if trained teachers leave for other jobs. Younger workers in a variety of occupations change jobs more frequently than their older counterparts (National Research Council, 1999). However, compared with other types of professionals, a higher proportion of teachers leave their positions each year. In response to surveys conducted in the mid-1990s, teachers indicated that, among the reasons they left their positions—including retirement, layoffs, and family reasons—dissatisfaction was one of the most important. Mathematics and science teachers reported more frequently than other teachers that job dissatisfaction was the reason they left their jobs. And, among teachers who left because of job dissatisfaction, mathematics and science teachers reported more frequently than other teachers that they left because of "poor administrative support" (Ingersoll, 2003, p. 7). The surveys defined "poor administrative support" as including a lack of recognition and support from administration and a lack of resources and material and equipment for the classroom.

Some research indicates that teachers do not respond to sustained professional development by taking their new knowledge and skills to other schools, but rather by staying and creating new benefits where they are. One study found that schools that provide more support to new teachers, including such professional development activities as induction and mentoring, have lower turnover rates (Ingersoll, 2003, p. 8). In addition, some researchers argue that, although professional development expends resources (time, money, supplies), it also creates new human and social resources (Gamoran et al., 2003, p. 28).

Gamoran and others studied six sites where teachers and educational researchers collaborated to reform science and mathematics teaching, focusing on teaching for understanding. "Teaching for understanding" was defined as including a focus on student thinking, attention to powerful scientific ideas, and the development of equitable classroom learning communities. Gamoran and colleagues found that, although the educational researchers provided an infusion of expertise from outside each of the six school sites, the professional development created in collaboration with the local schools had its greatest impact in supporting local teachers in developing their own communities. These school-based teacher communities, in turn, not only supported teachers in improving their teaching practices, but also helped them create new resources, such as new curricula. The teaching communities that developed, with their new leaders, succeeded in obtaining additional resources (such as shared teacher planning time) from within the schools and districts (Gamoran et al., 2003) and also from outside of them. Although the time frame of the study prevented analysis of whether the teacher communities were sustained over time, the results suggest that school districts can use focused professional development as a way to create strong teaching communities with the potential to support continued improvement in laboratory teaching and learning.

Scheduling Laboratory Teaching and Learning

Currently, most schools are designed to support teaching that follows predictable routines and schedules (Gamoran, 2004). Administrators allocate time, like other resources, as a way to support teachers in carrying out these routines. However, several types of inflexible scheduling may discourage effective laboratory experiences, including (a) limits on teacher planning time, (b) limits on teacher setup and cleanup time, and (c) limits on time for laboratory experiences.

Shared teacher planning time may be a critical support for improved laboratory teaching, because of the unique nature of laboratory education. As we have discussed, teachers face an ongoing tension between allowing students greater autonomy in the laboratory and guiding them toward accepted scientific knowledge. They also face uncertainty about how many variables students should struggle with and how much to narrow the context and procedures of the investigation. When one college physics professor taught a high school physics class, he struggled with uncertainty about how to respond to students' ideas about the phenomena they encountered, particularly when their findings contradicted accepted scientific principles (Hammer, 1997). In a case study of his experience, this professor called for reducing science teachers' class loads so they have more time to reflect on and improve their own practice.

A supportive school administration could help teachers overcome their isolation and learn from each other by providing time and space to reflect on their laboratory teaching and on student learning in the company of colleagues (Gamoran, 2004). In this approach, school administrators recognize that leadership for improved teaching and learning is distributed throughout the school and district and does not rest on traditional hierarchies.

In 2000, according to a nationally representative survey of science teachers, most school administrators provided inadequate time for shared planning and reflection to improve instruction. When asked whether they had time during the regular school week to work with colleagues on the curriculum and teaching, 69 percent of high school teachers disagreed and 4 percent had no opinion, leaving only 28 percent who agreed. However, 66 percent of teachers indicated that they regularly shared ideas and materials with their colleagues, perhaps indicating that they do so on their own time, outside school hours (Hudson et al., 2002). Only 11 percent of responding teachers indicated that science teachers in their school regularly observed other science teachers. Among teachers who acted as heads of science departments, 21 percent indicated that the lack of opportunities for teachers to share ideas was a serious problem for science instruction (Smith et al., 2002).

Time constraints can also discourage teachers from the challenges of setting up and testing laboratory equipment and materials. Associations of science teachers have taken differing positions on how administrators can best support teachers in preparing for and cleaning up after laboratory experiences. The American Association of Physics Teachers (AAPT) suggests that physics teachers should be required to teach no more than 275 instructional minutes per day. Many schools schedule eight 40- to 55-minute class periods, so that following the AAPT guidelines would allow physics teachers two preparation periods. The guidelines also call on administrators to schedule no more than 125 students per teacher per day, if the teacher is teaching only physics (the same laboratory activity taught several times may not require preparation) and no more than 100 students per teacher per day if the teacher is teaching both chemistry and physics, requiring more preparation time (American Association of Physics Teachers, 2002). The guidelines note that simply maintaining the laboratory requires at least one class period per day, and, if schools will not provide teachers with that time, they suggest that those schools either employ laboratory technicians or obtain student help.

The National Science Teachers Association takes a slightly different position, suggesting that administrators provide teachers with a competent paraprofessional. The paraprofessional would help with setup, cleanup, community contacts, searching for resources, and other types of support (National Science Teachers Association, 1990).

No national survey data are available to indicate whether science teachers receive adequate preparation time or assistance from trained laboratory technicians. Some individual teachers told our committee that they did not have adequate preparation and cleanup time.

Finally, adequate time is essential for student learning in laboratory experiences. On the basis of a review of the available research, Lunetta (1998, p. 253) suggests that, for students, "time should be provided for engaging students in driving questions, for team planning, for feedback about the nature and meaning of data, and for discussion of the implications of findings," and laboratory journals "should provide opportunities for individual students to reflect upon and clarify their own observations, hypotheses, conceptions."

School administrators can take several approaches to providing time for this type of ongoing discussion and reflection that supports student learning during laboratory experiences. Block scheduling is one approach schools have used to provide longer periods of time for laboratory activities and discussion. In this approach classes meet every other day for longer blocks of about 90-100 minutes, instead of every day for 40 or 45 minutes. However, an analysis of national survey data indicates that teachers in block schedules do not incorporate more laboratory experiences into their instruction (Smith, 2004). In addition, there is little research on whether use of block scheduling influences teachers' instruction or enhances student learning.

In another approach, schools can schedule science classes for double periods to allow more time for both carrying out investigations and reflecting on the meaning of those investigations. In an ideal world, administrators would provide adequate laboratory space and time to allow students to continue investigations over several weeks or months, and they would also provide time for students to work outside regular school hours. One study found that, when laboratories were easily accessible, 14- and 15-year-old students who used the facilities during their free time reported increased interest in academics and took advanced science courses (Henderson and Mapp, 2002).

SUMMARY

Teachers play a critical role in leading laboratory experiences in ways that support student learning. However, the undergraduate education of future science teachers does not currently prepare them for effective laboratory teaching. Undergraduate science departments rarely provide future science teachers with laboratory experiences that follow the design principles derived from recent research—integrated into the flow of instruction, focused on clear learning goals, aimed at the learning of science content and science process, with ongoing opportunities for reflection and discussion. Once on the job, science teachers have few opportunities to improve their laboratory teaching. Professional development opportunities for science teachers are limited in quality, availability, and scope and place little emphasis on laboratory instruction. Further research is needed to inform design of laboratory-focused teacher professional development that can support teachers in improving laboratory instruction. In addition, few high school teachers have access to curricula that integrate laboratory experiences into the stream of instruction

The organization and structure of most high schools impede teachers' and administrators' ongoing learning about science instruction and the implementation of quality laboratory experiences. Administrators who take a more flexible approach can support effective laboratory teaching by providing teachers with adequate time and space for ongoing professional development and shared lesson planning.

Improving high school science teachers' capacity to lead laboratory experiences effectively is critical to advancing the educational goals of these experiences. This would require both a major changes in undergraduate science education, including provision of a range of effective laboratory experiences for future teachers, and developing more comprehensive systems of support for teachers.

REFERENCES

American Association of Physics Teachers. (2002). *AAPT guidelines for high school physics programs*. Washington, DC: Author.

Anderson, C., Sheldon, T., and Dubay, J. (1990). The effects of instruction on college nonmajors' conceptions of respiration and photosynthesis. *Journal of Research in Science Teaching, 27,* 761-776.

Baumgartner, E. (2004). Synergy research and knowledge integration. In M.C. Linn, E.A. Davis, and P. Bell (Eds.), *Internet environments for science education*. Mahwah, NJ: Lawrence Earlbaum.

Bayer Corporation. (2004). *Bayer facts of science education 2004: Are the nation's colleges adequately preparing elementary schoolteachers of tomorrow to teach science?* Available at: http://www.bayerus.com/msms/news/facts.cfm?mode=detailandid-survey04 [accessed Dec. 2004].

Bell, P. (2004). *The school science laboratory: Considerations of learning, technology, and scientific practice.* Paper prepared for the Committee on High School Science Laboratories: Role and Vision, July 12-13, National Research Council, Washington, DC. Available at: http://www7.nationalacademies.org/bose/July_12-13_2004_High_School_Labs_Meeting_Agenda.html.

Brown, A.L., and Campione, J.C. (1998). Designing a community of young learners: Theoretical and practical lessons. In N.M. Lambert and B.L. McComs (Eds.), *How students learn: Reforming schools through learner-centered education* (pp. 153-186). Washington, DC: American Psychological Association.

Bruner, J. (1996). *The culture of education.* Cambridge, MA: Harvard University Press.

Catley, K. (2004). *How do teachers work and learn—specifically related to labs.* Presentation to the Committee on High School Science Laboratories: Role and Vision, June 3-4, National Research Council, Washington, DC. Available at: http://www7.nationalacademies.org/bose/June_3-4_2004_High_School_Labs_Meeting_Agenda.html [accessed Oct. 2004].

Chaney, B. (1995). *Student outcomes and the professional preparation of eighth-grade teachers in science and mathematics: NSF/NELS.* Rockville, MD: Westat.

Clark, R.L., Clough, M.P., and Berg, C.A. (2000). Modifying cookbook labs. *Science Teacher* (October), 40-43.

Darling-Hammond, L., Berry, B., and Thoreson, A. (2001). Does teacher certification matter? Evaluating the evidence. *Educational Evaluation and Policy Analysis, 23*(1), 57-77.

Deng, Z. (2001). The distinction between key ideas in teaching school physics and key ideas in the discipline of physics. *Science Education, 85*(3), 263-278.

DeSimone, L.M., Garet, M., Birman, B., Porter, A., and Yoon, K. (2003). Improving teachers' in-service professional development in mathematics and science: The role of postsecondary institutions. *Educational Policy, 17*(5), 613-649. Abstract available at: http://epx.sagepub.com/cgi/content/abstract/17/5/613 [accessed May 2005].

DeSimone, L.M., Porter, A.S., Garet, M.S., Yoon, K.S., and Birman, B. (2002). Effects of professional development on teachers' instruction: Results from a three-year longitudinal study. *Educational Evaluation and Policy Analysis, 24*(2), 81-112.

Driver, R. (1995). Constructivist approaches to science teaching. In L.P. Steffe and J. Gale (Eds.), *Constructivism in education.* Hillsdale, NJ: Lawrence Erlbaum.

Duschl, R. (1983). The elementary level science methods course: Breeding ground of an apprehension toward science? *Journal of Research in Science Teaching, 20,* 745-754.

Ferguson, R. (1998). Can schools narrow the black-white test score gap? In C. Jencks and M. Phillips (Eds.), *The black-white test score gap.* Washington, DC: Brookings Institution.

Fred Hutchinson Cancer Research Center. (2003). *Welcome to the Science Education Partnership.* Seattle: Author. Available at: http://www.fhcrc.org/education/sep/ [accessed Feb. 2005].

Gallagher, J. (1991). Prospective and practicing secondary school science teachers' knowledge and beliefs about the philosophy of science. *Science Education, 75,* 121-133.

Gamoran, A. (2004). *Organizational conditions that support inquiry in high school science instruction*. Presentation to the Committee on High School Science Laboratories: Role and Vision, June 3-4, National Research Council, Washington, DC. Available at: http://www7.nationalacademies.org/bose/June_3-4_2004_High_School_Labs_Meeting_Agenda.html [accessed May 2005].

Gamoran, A., Anderson, C.W., Quiroz, P.A., Seceda, W.G., Williams, T., and Ashmann, S. (2003). *Transforming teaching in math and science: How schools and districts can support change.* New York: Teachers College Press.

Gess-Newsome, J., and Lederman, N. (1993). Pre-service biology teachers' knowledge structures as a function of professional teacher education: A year-long assessment. *Science Education, 77*(1), 25-46.

Gitomer, D.H., and Duschl, R.A. (1998). Emerging issues and practices in science assessment. In B.J. Fraser and K.G. Tobin (Eds.), *International handbook of science education* (pp. 791-810). London, England: Kluwer Academic.

Glagovich, N., and Swierczynski, A. (2004). Teaching failure in the laboratory. *Journal of College Science Teaching, 33*(6).

Goldhaber, D.D. (2002). The mystery of good teaching: Surveying the evidence on student achievement and teachers' characteristics. *Education Next, 2*(1), 50-55. Available at: http://www.educationnext.org/20021/50.html [accessed Feb. 2005].

Goldhaber, D.D., and Brewer, D.J. (1997). Evaluating the effect of teacher degree level on educational performance. In W. Fowler (Ed.), *Development in school finance, 1996.* (ED 409-634.) Washington, DC: U.S. Department of Education, National Center for Education Statistics.

Goldhaber, D.D., and Brewer, D.J. (2001). Evaluating the evidence on teacher certification: A rejoinder. *Educational Evaluation and Policy Analysis, 23*(1), 79-86.

Goldhaber, D.D., Brewer, D.J., and Anderson, D. (1999). A three-way error components analysis of educational productivity. *Education Economics, 7*(3), 199-208.

Hammer, D. (1997). Discovery learning and discovery teaching. *Cognition and Instruction, 15*(4), 485-529.

Hanusek, E., Kain, J., and Rivkin, S. (1999). *Do higher salaries buy better teachers?* (Working Paper No. 7082.) Cambridge, MA: National Bureau of Economic Research.

Harlen, W. (2000). *Respecting children's own ideas.* New York: City College Workshop Center.

Harlen, W. (2001). *Primary science: Taking the plunge.* Portsmouth, NH: Heinemann.

Hegarty-Hazel, E. (1990). Life in science laboratory classrooms at the tertiary level. In E. Hegarty-Hazel (Ed.), *The student laboratory and the curriculum* (pp. 357-382). London, England: Routledge.

Hein, G.E., and Price, S. (1994). *Active assessment for active learning.* Portsmouth, NH: Heinemann.

Henderson, A.T., and Mapp, K.L. (2002). *A new wave of evidence—The impact of school, family, and community connections in student achievement.* National Center for Family and Community Connections with Schools. Austin, TX: Southwest Educational Development Laboratory. Available at: http://www.sedl.org/connections/research-syntheses.html [accessed May 2005].

Hilosky, A., Sutman, F., and Schmuckler, J. (1998). Is laboratory-based instruction in beginning college-level chemistry worth the effort and expense? *Journal of Chemical Education, 75*(1), 100-104.

Hirsch, E., Koppich, J.E., and Knapp, M.S. (2001). *Revisiting what states are doing to improve the quality of teaching: An update on patterns and trends.* (Working paper prepared in collaboration with the National Conference of State Legislatures.) Seattle: University of Washington, Center for the Study of Teaching and Policy.

Hudson, S.B., McMahon, K.C., and Overstreet, C.M. (2002). *The 2000 National Survey of Science and Mathematics Education: Compendium of tables.* Chapel Hill, NC: Horizon Research.

Ingersoll, R. (2003). Is there a shortage among mathematics and science teachers? *Science Educator, 12*(1), 1-9.

Javonovic, J., and King, S.S. (1998). Boys and girls in the performance-based classroom: Who's doing the performing? *American Educational Research Journal 35*(3), 477-496.

Kennedy, M., Ball, D., McDiarmid, G.W., and Schmidt, W. (1991). *A study package for examining and tracking changes in teachers' knowledge.* East Lansing, MI: National Center for Research in Teacher Education.

Khalic, A., and Lederman, N. (2000). Improving science teachers' conceptions of nature of science: A critical review of the literature. *International Journal of Science Education 22*(7), 665-701.

Lederman, N.G. (2004). *Laboratory experiences and their role in science education.* Presentation to the Committee on High School Science Laboratories: Role and Vision, July 12-13, National Research Council, Washington, DC. Available at: http://www7.nationalacademies.org/bose/July_12-13_2004_High_School_Labs_Meeting_Agenda.html [accessed May 2005].

Lee, O. (2002). Equity for linguistically and culturally diverse students in science education. A research agenda. *Teachers College Record, 105*(3), 465-489.

Lee, O., and Fradd, S.H. (1998). Science for all, including students from non-English-language backgrounds. *Educational Researcher, 27,* 12-21.

Linn, M.C. (1995). Designing computer learning environments for engineering and computer science: The scaffolded knowledge integration framework. *Journal of Science Education and Technology, 4*(2), 103-126.

Linn, M.C. (1997). The role of the laboratory in science learning. *Elementary School Journal, 97*(4), 401-417.

Linn, M.C., Davis, E.A., and Bell, P. (2004). *Internet environments for science education.* Mahwah, NJ: Lawrence Erlbaum.

Loucks-Horsley, S., Love, N., Stiles, K.E., Mundry, S., and Hewson, P.W. (2003). *Designing professional development for teachers of science and mathematics.* Thousand Oaks, CA: Corwin Press.

Lunetta, V.N. (1998). The school science laboratory: Historical perspectives and contexts for contemporary teaching. In B.J. Fraser and K.G. Tobin (Eds.), *International handbook of science education* (pp. 249-262). London, England: Kluwer Academic.

Lynch, S., Kuipers, J., Pike, C., and Szeze, M. (in press). Examining the effects of a highly rated curriculum unit on diverse students: Results from a planning grant. *Journal of Research in Science Teaching.*

Maienschein, J. (2004). *Laboratories in science education: Understanding the history and nature of science.* Presentation to the Committee on High School Science Laboratories: Role and Vision, June 3-4, National Research Council, Washington, DC. Available at: http://www7.nationalacademies.org/bose/June_3-4_2004_High_School_Labs_Meeting_Agenda.html [accessed May 2005].

McComas, W.F., and Colburn, A.I. (1995). Laboratory learning: Addressing a neglected dimension of science teacher education. *Journal of Science Teacher Education, 6*(2), 120-124.

McDiarmid, G.W. (1994). The arts and science as preparation for teaching. In K. Howey and N. Zimpher (Eds.), *Faculty development for improving teacher preparation* (pp. 99-138). Reston, VA: Association of Teacher Educators.

McDiarmid, G.S., Ball, D.L., and Anderson, C.W. (1989). Why staying ahead one chapter doesn't really work: Subject-specific pedagogy. In M.D. Reynolds (Ed.), *Knowledge base for the beginning teacher.* New York: Pergamon.

Millar, R. (2004). *The role of practical work in the teaching and learning of science.* Paper prepared for the Committee on High School Science Laboratories: Role and Vision, July 12-13, National Research Council, Washington, DC. Available at: http://www7.nationalacademies.org/bose/June_3-4_2004_High_School_Labs_Meeting_Agenda.html [accessed May 2005].

Millar, R., and Driver, R. (1987). Beyond process. *Studies in Science Education, 14,* 33-62.

Minstrell, J., and van Zee, E.H. (2003). Using questioning to assess and foster student thinking. In J.M. Atkin and J.E. Coffey, *Everyday assessment in the science classroom* (pp. 61-74). Arlington, VA: National Science Teachers Association.

Mortimer, E., and Scott, P. (2003). *Meaning making in secondary science classrooms.* Philadelphia: Open University Press.

National Center for Education Statistics. (2004). *Qualifications of the public school teacher workforce: Prevalence of out-of-field teaching 1987-88 to 1999-2000.* Statistical analysis report. Washington, DC: Author.

National Research Council. (1990). *Fulfilling the promise: Biology education in the nation's schools.* Committee on High School Biology Education, Commission on Life Sciences. Washington, DC: National Academy Press.

National Research Council. (1999). *The changing nature of work: Implications for occupational analysis.* Committee on Techniques for the Enhancement of Human Performance: Occupational Analysis. Washington, DC: National Academy Press.

National Research Council. (2001a). *Educating teachers of science, mathematics, and technology.* Committee on Science and Mathematics Teacher Preparation, Center for Education. Washington, DC: National Academy Press.

National Research Council. (2001b). *Classroom assessment and the national science education standards.* Committee on Classroom Assessment and the National Science Education Standards, J.M. Atkin, P. Black, and J. Coffey (Eds.). Center for Education. Washington, DC: National Academy Press.

National Science Teachers Association. (1990). *NSTA position statement: Laboratory science.* Available at: http://www.nsta.org/positionstatementandpsid=16 [accessed Oct. 2004].

Olsen, T.P., Hewson, P.W., and Lyons, L. (1996). Preordained science and student autonomy: The nature of laboratory tasks in physics classrooms. *International Journal of Science Education, 18*(7), 775-790.

Pomeroy, D. (1993). Implications of teachers' beliefs about the nature of science: Comparisons of the beliefs of scientists, secondary science teachers, and elementary science teachers. *Science Education, 77,* 261-278.

Priestley, W., Priestley, H., and Schmuckler, J. (1997). *The impact of longer term intervention on reforming the approaches to instructions in chemistry by urban teachers of physical and life sciences at the secondary school level.* Paper presented at the National Association for Research in Science Teaching meeting, March 23, Chicago, IL.

Sanders, M. (1993). Erroneous ideas about respiration: The teacher factor. *Journal of Research in Science Teaching, 30,* 919-934.

Sanders, W.L., and Rivers, J.C. (1996). *Cumulative and residual effects of teachers on future student academic achievement.* Knoxville: University of Tennessee Value-Added Research and Assessment Center.

Schwartz, R., and Lederman, N. (2002). It's the nature of the beast: The influence of knowledge and intentions on learning and teaching nature of science. *Journal of Research in Science Teaching, 39*(3), 205-236.

Shulman, L.S. (1986). Those who understand: Knowledge growth in teaching. *Educational Researcher, 15,* 4-14.

Slotta, J.D. (2004). The web-based inquiry science environment (WISE): Scaffolding knowledge integration in the science classroom. In M.C. Linn, E.A. Davis, and P. Bell (Eds.), *Internet environments for science education.* Mahwah, NJ: Lawrence Earlbaum.

Smith, P.S., Banilower, E.R., McMahon, K.C., and Weiss, I.R. (2002). *The National Survey of Science and Mathematics Education: Trends from 1977 to 2000.* Chapel Hill, NC: Horizon Research. Available at: http://www.horizon-research.com/reports/2002/2000survey/trends.php [accessed May 2005].

Smith, S. (2004). *High school science laboratories. Data from the 2000 National Survey of Science and Mathematics Education.* Presentation to the NRC Committee on High School Science Laboratories, March 29, Washington, DC. Available at: http://www7.nationalacademies.org/bose/March_29-30_2004_High_School_Labs_Meeting_Agenda.html [accessed Oct. 2005].

Songer, C., and Mintzes, J. (1994). Understanding cellular respiration: An analysis of conceptual change in college biology. *Journal of Research in Science Teaching, 31,* 621-637.

Supovitz, J.A., Mayer, D.P., and Kahle, J. (2000). Promoting inquiry-based instructional practice: The longitudinal impact of professional development in the context of systemic reform. *Educational Policy, 14*(3), 331-356.

Supovitz, J.A., and Turner, H.M. (2000). The effects of professional development on science teaching practices and classroom culture. *Journal of Research on Science Teaching, 37,* 963-980.

Sutman, F.X., Schmuckler, J.S., Hilosky, A.B., Priestly, H.S., and Priestly, W.J. (1996). *Seeking more effective outcomes from science laboratory experiences (Grades 7-14): Six companion studies.* Paper presented at the annual meeting of the National Association for Research in Science Teaching, April, St. Louis, MO.

Tobin, K.G. (2004). *Culturally adaptive teaching and learning science in labs.* Presentation to the Committee on High School Science Laboratories: Role and Vision, July 12-13, National Research Council, Washington, DC. Available at: http://www7.nationalacademies.org/bose/KTobin_71204_HSLabs_Mtg.pdf [accessed August 2005].

Trumbull, D., and Kerr, P. (1993). University researchers' inchoate critiques of science teaching: Implications for the content of pre-service science teacher education. *Science Education, 77*(3), 301-317.

Tushnet, N.C., Millsap, M.A., Noraini, A., Brigham, N., Cooley, E., Elliott, J., Johnston, K., Martinez, A., Nierenberg, M., and Rosenblum, S. (2000). *Final report on the evaluation of the National Science Foundation's Instructional Materials Development Program.* Arlington, VA: National Science Foundation.

U.S. Department of Education. (2000). *Before it's too late: A report to the nation from the national commission on mathematics and science teaching for the 21st century.* Washington, DC: Author.

U.S. Department of Education. (2004). *The condition of education.* Washington, DC: Author. Available at: http://www.nces.ed.gov/programs/coe/2004/section4/indicator24.asp [accessed Feb. 2005].

U.S. Department of Energy. (2004). *The laboratory science teacher professional development program.* Washington, DC: Author. Available at: http://www.scied.science.doe.gov/scied/LSTPD/about.htm [accessed Feb. 2005].

van Zee, E., and Minstrell, J. (1997). Using questioning to guide student thinking. *Journal of the Learning Sciences, 6*(2), 227-269.

Volkmann, M., and Abell, S. (2003). Rethinking laboratories. *Science Teacher,* September, 38-41.

Weiss, I.R., Pasley, J.D., Smith, P.S., Banilower, E.R., and Heck, D.J. (2003). *Looking inside the classroom: A study of K-12 mathematics and science education in the United States.* Chapel Hill, NC: Horizon Research.

Westbrook, S., and Marek, E. (1992). A cross-age study of student understanding of the concept of homeostasis. *Journal of Research in Science Teaching, 29*, 51-61.

Williams, M., Linn, M.C., Ammon, P., and Gearhart, M. (2004). Learning to teach inquiry science in a technology-based environment: A case study. *Journal of Science Education and Technology, 13*(2), 189-206.

Windschitl, M. (2004). *What types of knowledge do teachers use to engage learners in "doing science"? Rethinking the continuum of preparation and professional development for secondary science educators.* Paper prepared for the Committee on High School Science Laboratories: Role and Vision, June 3-4, National Research Council, Washington, DC. Available at: http://www7.nationalacademies.org/bose/June_3-4_2004_High_School_Labs_Meeting_Agenda.html [accessed May 2005].

Wright, S.P., Horn, S., and Sanders, W. (1997). Teacher and classroom context effects on student achievement: Implications for teacher evaluation. *Journal of Personnel Evaluation in Education, 11*(1), 57-67.

Zahopoulos, C. (2003). Retired scientists and engineers: Providing in-classroom support to K-12 science teachers. In D.G. Haase, B.S. Wojnowski, and S.K. Schulze (Eds.), *Proceedings of the Conference on K-12 Outreach from University Science Departments.* Raleigh: Science House, North Carolina State University.

6

Facilities, Equipment, and Safety

Key Points

- *The design of space for laboratory experiences that follow the principles developed in this report would allow for flexible use of space and furnishings, combining features of traditional laboratories and classrooms.*
- *In budgeting for laboratories, schools must consider the ongoing costs of equipment and supplies as well as the costs of building facilities.*
- *Adequate facilities, equipment, and supplies for laboratory experiences are inequitably distributed.*
- *Maintaining student safety during laboratory experiences is a critical concern, but little systematic information is available about safety problems and solutions.*

In this chapter we discuss the challenges of providing appropriate physical space for laboratory experiences, including attention to equipment and supplies. In the first section we discuss the considerations regarding learning and teaching that must inform the design of laboratory space. We consider the complexities of budgeting for laboratory facilities, including options when resources are scarce. In the second section, we review disparities in the distribution of laboratory facilities, equipment, and supplies. In the final section we discuss laboratory safety, including attention to liability, standards of care for student safety, and current patterns of safety enforcement.

PROVIDING FACILITIES, EQUIPMENT, AND SUPPLIES

In response to growing enrollments and the deterioration of an older generation of buildings, school districts across the nation are involved in a wave of construction and renovation. A comprehensive survey conducted by the General Accounting Office in 1996 revealed that many existing school buildings were in need of reconstruction or renovation. At that time, one-third of schools across the nation needed either extensive renovation or reconstruction, while another third had at least one major structural flaw, such as a leaky roof, an outdated electrical system, or dysfunctional plumbing (U.S. General Accounting Office, 1996).

On average, public elementary and secondary schools across the nation are devoting an increasing share of their budgets—from 10 percent in 1989-1990 to 14 percent in 2002-2002—to capital investments (National Center for Education Statistics, 2004b). Trend data from an annual mail and telephone survey of school district chief business officers indicate that planned and completed school construction spending nearly doubled over the past decade, increasing from $10.7 billion in 1994 to $28.6 billion in 2003 (Agron, 2003). About 61 percent of these expenditures was for new construction, and 39 percent was for additions or renovations to existing buildings. Another recent survey found that spending on school construction projects to be completed in 2003 totaled $19.7 billion, with 64 percent of the total dedicated to new construction, 21 percent for additions to existing buildings, and 14 percent for renovations of existing structures (Abramson, 2004). Respondents to the second survey indicated that 41 percent of expenditures for projects to be completed in 2003 were for high schools.[1] They indicated that 100 percent of new high schools and 92 percent of new middle schools would include science laboratories (Abramson, 2004). Laboratory facilities were included as part of additions to existing schools much less frequently (in about 18 percent of high school projects and 8 percent of middle school projects).

Laboratory Design and Student Learning

Specialized space for carrying out laboratory experiences can be incorporated into the initial design of a school or added or enhanced through

---

[1] Neither survey provides information on sampling design or response rate.

reconstruction and renovation. For new construction, reconstruction, or renovation, a critical consideration in creating space for laboratory experiences is how the design can best support science learning and teaching.

Over the past decade, there has been little research examining the relationship between physical laboratory spaces and student learning. The few studies available suggest that laboratory facilities influence teaching and student learning in poorly understood ways. As part of a comprehensive evaluation of Australia's science education curriculum, the government surveyed teachers about laboratory facilities and students' perceptions of their learning environments. The results suggested that active forms of learning were associated with better science facilities (Ainley, 1978, 1990). U.S. studies conducted in the late 1960s and early 1970s found that inquiry teaching methods were more frequent in spaces with combined classroom and laboratory facilities, compared with teaching in spaces where the classroom and laboratory are separate (Englehardt, 1968).

One researcher in Israel considered the history of the transformation of chemistry laboratories (in Europe, the United States, and elsewhere) from fixed benches with rows of reagent bottles to more open, flexible layouts that allowed better communication and collaboration between teachers and students (Arzi, 1998). She concluded, on the basis of this history and other research, that not only are science teachers influenced by space design, but they also influence those designs (Arzi, 1998). More recently, Henderson, Fisher, and Fraser found a significant positive correlation between students' perceptions of the material environment and students' attitude toward both laboratory experiences and science class (Henderson, Fisher, and Fraser, 2000). Students' perceptions of the material environment were determined using the Science Laboratory Environment Inventory (see Chapter 3).

A case study of one high school illustrates how the availability and quality of laboratory facilities may influence the availability and quality of effective teachers. When parents in this poor inner-city school found that one reason the school could not recruit a science teacher was a lack of laboratories, they organized to demand improvements from school district administrators. They won a $5 million rehabilitation program that included new science laboratories (Henderson and Mapp, 2002, pp. 58, 128).

Because of the expense of constructing or renovating laboratory space, the design should be future-oriented, supporting a vision of the science program over a decade or more. The first step in designing laboratory space is to develop such a long-term vision for the school science curriculum. The school science supervisor, along with curriculum coordinators, other science teachers, administrators, and state and local experts, often play important roles in developing this vision (Biehle, Motz, and West, 1999).

While the design of particular facilities will vary depending on the local science curriculum, available resources, and building codes, all school laboratory facilities should provide space for shared teacher planning, space for preparation of investigations, and secure storage for laboratory supplies as well as space for student activities and teacher demonstrations. In addition, past studies (Novak, 1972; Shepherd, 1974) and current laboratory design experts (Lidsky, 2004) agree that laboratory designs should emphasize flexible use of space and furnishings to support integration of laboratory experiences with other forms of science instruction.

Combined laboratory-classrooms can support effective laboratory experiences by providing movable benches and chairs, movable walls, peripheral or central location of facilities, wireless Internet connections and trolleys for computers, fume hoods, or other equipment. These flexible furnishings allow students to move seamlessly from carrying out laboratory activities on the benches to small-group or whole-class discussions that help them make meaning from their activities. Integrated laboratory-classrooms that provide space for long-term student projects or cumulative portfolios support the full range of laboratory experiences, allowing students to experience more of the activities of real scientists. Forward-looking laboratory designs maximize use of natural sunlight and provide easy access to outdoor science facilities. See Figures 6-1 and 6-2 for examples of laboratory-classrooms with flexible designs.

Designing school laboratory spaces to accommodate multiple science disciplines could provide both educational and practical benefits. First, because undergraduate science education, like science itself, is becoming more interdisciplinary, a National Research Council committee has recommended making undergraduate laboratory courses as interdisciplinary as possible (National Research Council, 2003). High school laboratory facilities that could accommodate interdisciplinary investigations would help prepare students for such undergraduate laboratory courses. Second, high school students enroll in a wide variety of science courses (National Center for Education Statistics, 2004).[2] It may be more cost-effective to provide this variety with a few laboratory classrooms that can accommodate multiple disciplines than by constructing discipline-specific laboratory classrooms that remain unused at times.

The committee was unable to locate any systematic national data on the extent to which current high school science laboratory spaces incorporate any of the aspects of flexibility described above. No systematic information was available on the extent to which high school science classrooms may be

---

[2]For example, in California, among the 74,000 high school science classes offered during the 2002-2003 school year, the largest group (37 percent) was in general science, followed by life science classes (27 percent). Classes in other science subjects made up much smaller shares of the total, including chemistry (9 percent), integrated science (7 percent), and physics (4 percent) (California Commission on Teacher Credentialling, 2004).

designed to allow for easy movement from laboratory work to group discussions or lectures and/or to accommodate multiple science disciplines. For example, almost no information was available on the fraction of high schools that include combined laboratory-classroom space instead of separate laboratory rooms. In 1999, two teachers' associations—the National Science Teachers Association and the International Technology Education Association—mailed a survey to their members and received about 2,000 responses (LabPlan, 2004). Among the 900 National Science Teachers Association members who responded, over three-fourths indicated they taught in combined laboratory-classrooms. Among the 1,200 responding International Technol-

FIGURE 6-1 Laboratory classroom set up for group laboratory work and teacher demonstration or mini-lecture.
SOURCE: Lidsky (2004).

FIGURE 6-2 Laboratory classroom set up for small-group investigations at central benches and individual activities at side benches.
SOURCE: Lidsky (2004).

ogy Education Association members, who taught drafting, technology education, and manufacturing courses, just under half taught in a combined laboratory-classroom and one-quarter taught in a combined laboratory–production classroom (LabPlan, 2004).

Budgeting for Laboratory Facilities, Equipment, and Supplies

Because laboratories require space for student activities, shared teacher planning, teacher demonstrations, student discussions, and safe storage of chemicals, along with specialized furnishings (e.g., sinks, benches) and utilities (e.g., water, gas), they are more expensive to build and maintain than other types of school space. One recent guide to school science facilities indicates that "laboratory space is approximately twice as expensive to build and equip as classroom space." (Biehle et al., 1999, p. 56). According to one architect specializing in educational science laboratories, in 2004, the costs of laboratory space in New England ranged from $180 per square foot for general science and physics to $250 per square foot for chemistry and biology (Lidsky, 2004). At $250 per square foot, these laboratory costs are about 1.7 times more expensive than the costs of new high school space in New England, estimated at $148 per square foot in a recent survey (Abramson, 2004).

Daniel Gohl, principal of McKinley Technical High School in Washington, DC, pointed out that laboratories are effectively financed through two different budgets (Gohl, 2004). Although funds to plan, design, and build a new laboratory facility come from the school or district's capital budget, the supplies and equipment needed to use the laboratory space come out of the operating budget. In some cases, there may be enough capital budget to build a laboratory, but no funds are set aside in the operating budget to provide the equipment and supplies to use the laboratory over subsequent years. Gohl observed:

> It is not uncommon in jurisdictions throughout the country to find people who invest a tremendous amount of money in high tech [high-voltage alternating current] systems, great science labs, and then underfund them historically once they are built. It may be that there is no equipment, or it may be that they buy the equipment once and they don't buy the disposable materials every year in order to use them. There is no consensus as to how one budgets those resources into the foreseeable future.

A study in New York City supports Gohl's observations regarding budgeting for operational costs of labs (Schenk and Meeks, 1999). The New York State Regents exam has exerted pressure for high schools throughout the state, including those in New York City, to increase the number of laboratory courses offered. In New York City, 16 of the 18 schools surveyed increased the number of science classes requiring laboratory experiments between 1993-1994 and 1996-1997. In nine of the schools, the laboratory load at least doubled. The average increase in laboratory load between 1993-1994 and 1996-1997 was 90 percent.

Changes in the budget for laboratory materials and supplies were not commensurate with these increases in laboratory loads. For example, in one high school, although the number of laboratory sections tripled from 25 to 75, the school received only $300 more for materials and supplies. In the same time period, 7 of the 18 schools studied experienced a cut in their science budgets, and 5 of these schools simultaneously experienced increases in their laboratory load. For the nine schools that experienced an increase in their science budget, the budget increased 34 percent while the corresponding increase in laboratory load was 288 percent.

Designing Laboratory Experiences and Facilities when Resources are Scarce

When limited funds prevent schools from designing and constructing laboratory facilities in the school, there are alternative ways to provide students with effective laboratory experiences. Schools and teachers can arrange field trips with the help of local groups, such as the Audubon Naturalist Society, the local science museum, and the state department of natural resources.

Schools in rural areas may be able to obtain one of a growing number of mobile school laboratories (see Box 6-1). Laboratories on wheels can provide facilities, equipment, and trained teachers to rural students, and many of these laboratories also provide teacher professional development. However, because these projects typically rely on a variety of funding sources, including grants, they are not always sustainable. For example, Virginia Polytechnic University's Mobile Chemistry Laboratory, which relied on a combination of federal, corporate, private, and university funding, announced that operations would cease in early May of 2004 due to lack of funds. However, the National Science Foundation provided temporary funding to sustain the program through the 2004-2005 school year (Virginia Polytechnic Institute, 2004). In contrast, the Juniata College Science in Motion Program in Pennsylvania, initially funded by the National Science Foundation, has been sustained with state funding since 1997 (Mulfinger, 2004).

A few school districts and cities have found economies of scale by centralizing laboratory facilities in one location (this can be either an alternative to having laboratory facilities in every school or a supplement). For example, the Howard Hughes Medical Institute has supported new biotechnology laboratory facilities at Sterling High School in Loudoun County, Virginia, and a magnet program. Students across the county will use the laboratories at Sterling every other day, attending their home high schools for other courses and extracurricular activities (Helderman, 2004). In Tel Aviv, Israel, a centralized science facility performs a similar role, serving students from several schools with laboratory facilities and expert science laboratory teachers (Arzi, 1998). Students in Tel Aviv attend their home schools for other subjects and the science center for science. The Tel Aviv center has proven particularly effective in building teachers' knowledge and expertise for laboratory teaching, by providing a place for ongoing teacher collaboration, reflection, and improvement of instruction (see Figure 6-3).

BOX 6-1 Laboratories on Wheels

California: Teachers + Occidental = Partnership in Science (TOPS)
Available at: http://www.lalc.k12.ca.us/catalog/providers/172.html.

Colorado: Colorado State University Mobile Investigations
Available at: http://www.hhmi.org/news/csugia.html.

Delaware: Science Van Project—Science In Motion

Illinois:
Chicago State University Chemistry Van
Available at: http://members.tripod.com/~tyff/Outreach/chemvan.html.

University of Illinois at Urbana-Champaign Physics Van
Available at: http://van.hep.uiuc.edu/.

Northern Illinois University Frontier Physics
Available at: http://www.physics.niu.edu/~frontier/.

Indiana: Purdue University Instrument Van Project
Available at: http://www.chem.purdue.edu/cmobile/Chemobile%20%home%20page.htm.

New York: Marist College: Science on the Move
Available at: http://library.marist.edu/SOTM.

Pennsylvania: Science in Motion
Available at: http://www.science-in-motion.org/.

North Carolina: Science House Satellite Offices
Available at: http://www.science-house.org/info/satellite.html.

South Dakota: Science on the Move
Available at: http://www.camse.org/scienceonthemove/what_is_sotm.html.

West Virginia: Science on Wheels
Available at: http://www.marshall.edu/coe/toyota/.

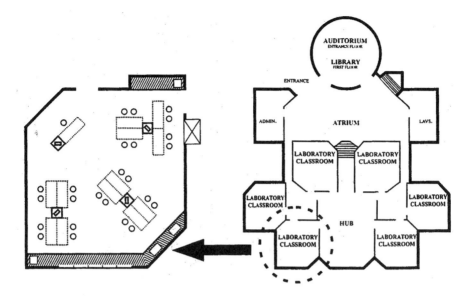

FIGURE 6-3 Schematic illustration of a laboratory-classroom and floor plan at HEMDA-Centre for Science Education in Tel Aviv, Israel.
SOURCE: Arzi (1998). Reprinted with permission.

DISPARITIES IN FACILITIES, EQUIPMENT, AND SUPPLIES

Disparities in Laboratory Facilities

Although well-designed flexible laboratory spaces can support effective laboratory experiences, access to such space is not available to all schools and students. Among science department heads surveyed in 2000, 21 percent indicated that facilities posed a serious problem for science instruction in their school (Smith, Banilower, McMahon, and Weiss, 2002). This represented an increase from 1993, when about 18 percent of heads of science departments indicated that facilities posed a serious problem.

In 1994, the U.S. General Accounting Office (GAO) surveyed a nationally representative sample of 10,000 schools in 5,000 school districts. This was the same sample used by the National Center for Education Statistics Schools and Staffing Survey administered by the Census Bureau. GAO mailed surveys to facilities directors and administrators in the school districts in which the sampled schools were located and received a 78 percent response.

TABLE 6-1 Percentage of Schools Reporting Inadequate Facilities by Proportion of Minority Students

	Percentage of Total			
Minority enrollment	50.5 or more	20.5-50.4	5.5-20.4	less than 5.5
Schools reporting inadequate facilities	49	43	39	39

SOURCE: U.S. General Accounting Office (1996, pp. 49-50).

Survey results were statistically adjusted to produce representative estimates at the national and state levels (U.S. General Accounting Office, 1996, p. 32). One survey question asked, "How well do this school's on-site buildings meet the functional requirements of the activities below?—very well, moderately well, somewhat well, not well at all." The list of activities included laboratory science. A total of 15 percent of respondents who were asked about high schools indicated that their laboratory facilities met functional requirements "not well at all." Specifically, they indicated that their facilities did not meet the following functional requirements for laboratory science: demonstration stations, student laboratory stations, safety equipment, and appropriate storage for chemicals and other supplies.

In its analysis of survey responses and school and student characteristics, GAO included responses about both elementary and secondary school buildings. The survey identified three trends. First, inadequate laboratory facilities varied by community type. The highest percentage of ill-equipped schools was in central cities, followed by urban fringes or large towns, and the smallest percentage of ill-equipped schools was in rural areas or small towns. Second, inadequate laboratory facilities varied by proportion of minority students, with less adequate laboratory facilities in schools with higher concentrations of minorities (see Table 6-1). Third, inadequate laboratories were associated with the proportion of students approved for free or reduced-price lunch, with less adequate facilities in schools with higher concentrations of students eligible for reduced-price meals (see Table 6-2).

More recent data regarding the adequacy of science facilities are available from a survey of school principals in New Jersey conducted in 2003 by Mark Schneider. Due to its focus on a single state, less careful design, and lower response rate, the results of this survey are less conclusive than the earlier GAO survey. In fall 2003, 1,300 principals who were members of the New Jersey Principals and Supervisors Association were sent surveys by email and fax. The response rate was about 20 percent. An analysis of the sample of respondents found that principals in New Jersey's poorest districts

TABLE 6-2 Percentage of Schools Reporting Inadequate Facilities by Proportion of Students Eligible for Free or Reduced-Price Lunch

	Percentage of Total			
Students eligible for free or reduced-price lunch	70 or more	40-70	20-40	Less than 20
Schools reporting inadequate facilities	50.3	49	38	20

SOURCE: U.S. General Accounting Office (1996, pp. 49-50).

were underrepresented in the sample (Schneider, 2004). In this survey, principals were not given any guidance to judge the adequacy of their laboratory facilities. The survey question asked whether specialized science facilities were "very adequate," "somewhat adequate," or "less than adequate" (p. 5). Despite these weaknesses, the survey responses are discussed here, as they are among the very few data available on the adequacy of laboratory facilities.

In response to this survey, principals in the poorest schools were more likely to find the facilities for science to be inadequate than principals in average or high-income schools. Specifically, around 30 percent of principals in poor schools indicated their science facilities were inadequate compared with less than 10 percent of principals in average or high-income schools (Schneider, 2004).

Surveys in three large cities with large concentrations of poor and minority students also reveal inadequate laboratory facilities and equipment. A survey conducted in New York City in the mid-1990s found that the few available laboratory rooms were in constant use, with teachers rotating in and out of classrooms. With no time to clean up broken glass or spills, student safety was compromised (Schenk and Meeks, 1999). In response to phone and mail surveys conducted in 2002, almost 60 percent of science teachers in Chicago and Washington, DC, reported either that their science laboratory facilities were somewhat inadequate or very inadequate to meet curriculum standards or that they had no science laboratory facilities at all (Schneider, 2002).

Disparities in Laboratory Equipment

Equipment necessary to safely conduct a variety of laboratory experiments is not available in all classrooms. In the 2000 survey of science education, high school teachers were asked about equipment used in their science classes. They responded that they frequently used electricity (in over 90 percent of classes), running water (over 90 percent of classes), gas for burners (in 72 percent of classes), and hoods or air hoses to remove dangerous fumes (in 56 percent of classes) (Smith et al., 2002). On average, teachers indicated that less than 10 percent of their science classes lacked needed access to electric outlets, running water, and gas for burners. However, teachers indicated that a larger share of science classes (26 percent in 1993 and 11 percent in 2000) would have benefited from hoods or air hoses that were unavailable.

A follow-up analysis of this national survey data revealed disparities in the availability of laboratory equipment and supplies. Teachers in schools with the highest concentrations of non-Asian minority students were more likely than teachers in other schools to indicate that fume hoods or air hoses were needed but not available (Banilower, Green, and Smith, 2004). Teachers in rural schools reported spending far less on equipment than teachers in urban or suburban areas, and equipment spending also varied by poverty and the ethnic composition of schools. The median of equipment spending was $4 per pupil per year among schools in the lowest poverty quartile, compared with $2 per pupil per year among schools in the highest poverty quartile (Banilower et al., 2004, p. 33). Equipment spending was also lower in schools with higher concentrations of non-Asian minorities. The median amount spent per pupil per year was $3 in schools with the lowest concentration of minorities and only $1 in schools with the highest concentration. For a typical high school of about 1,100 students, the median of annual spending on equipment was $3,126 in the lowest poverty schools and $1,935 in the highest poverty schools. Similarly, annual spending was $2,476 in schools with the fewest minorities and $1,928 in schools with the highest concentrations of minorities (Banilower et al., 2004, p. 32; see Table 6-3).

Disparities in Supplies

A study of high school biology in Israel found that "easy access to laboratory materials, living organisms, and chemicals has been influential in promoting laboratory work" (Tamir, 1976, cited in Lazarowitz and Tamir, 1994, p. 111). As in the case of laboratory facilities and equipment, such easy access is not equally distributed to all high schools and all high school students.

Nationally, the median of yearly expenditures per pupil for consumable science supplies was $3 in 2000, or $3,444 for a typical high school of about 1,100 students. However, the national average masks disparities in spending between rural schools, which have median science supply expenditures of only $994, and urban and rural schools, which have annual supply expenditures of $2,957 and $2,905, respectively (Banilower et al., 2004). In terms of annual spending on science supplies per pupil, schools with the lowest concentrations of poor students spent twice as much ($6) as the highest

TABLE 6-3 Spending on Laboratory Equipment (Median Amount Spent in 2000 by a Typical High School of 1,100 Students)

Demographic Category	Median Amount ($)	Standard Error
Overall	2,538	(253)
School location		
Urban	2,957	(464)
Rural	994	(292)
Suburban	2,905	(506)
School poverty		
1st quartile (lowest poverty)	3,126	(380)
2nd quartile	1,842	(335)
3rd quartile	2,758	(1,039)
4th quartile	1,928	(445)
Non-Asian minority percentage		
1st quartile (lowest percentage)	2,476	(383)
2nd quartile	2,372	(526)
3rd quartile	2,926	(812)
4th quartile (highest percentage)	1,928	(445)

SOURCE: Banilower, Green, and Smith (2004).

poverty schools ($3) on consumable supplies for science instruction. Annual spending per pupil was $5 in schools with the fewest non-Asian minorities and $3 in schools with the highest concentration of non-Asian minorities.

An earlier survey in New York City revealed that one South Bronx high school had no facilities designed specifically for laboratory investigations. This survey indicated that the average annual expenditure per pupil for laboratory supplies was merely $2.02 (range from $.93 to $3.31), well below the $16.79 per student per year spent by a suburban district on Long Island (Schenck and Meeks, 1999).

Lack of adequate supplies and access to those supplies can have severe effects on teaching and learning. A science teacher at an urban high school in a poor neighborhood of Washington, DC, told the committee that he teaches laboratories only every two weeks because of the challenge of obtaining and assembling supplies. The science supervisor of a rural school district in southwestern Virginia, speaking to the committee, described the challenge of teaching laboratory classes off a cart of equipment and supplies and teaching ecology in the school library.

Lack of available, accessible laboratory equipment and supplies forces some teachers to purchase these items out of their own pockets. In response to a 2000 survey, high school teachers indicated they spent an average of $55 per year of their own money for science classes (Smith et al., 2002, p.

62). A national survey by a trade association found that teachers spent an average of $589 of their own money on supplies in 2001, up from $448 in 1999 (Trejos, 2003). Recognizing this problem, in March 2002 President Bush signed into law an economic stimulus package that included an annual $250 deduction for teachers' personal expenditures on classroom supplies. In fall 2004, this tax deduction was extended for two years.

LABORATORY SAFETY

Questions about laboratory safety were not part of the committee's charge, yet safety issues emerged as a critical concern over the course of the study. This section provides a brief review of safety issues. Science teachers and schools have clear legal liability for the safety of students engaged in laboratory activities, and local, state, and federal regulations, codes, and policies provide clear specifications for ensuring student safety. The limited evidence available suggests that some U.S. high schools are not ready to provide safe laboratory activities.

Liability for Student Safety

As defined by U.S. courts today, "negligence" is conduct that falls below a standard of care established by law or profession to protect others from an unreasonable risk of harm, or the failure to exercise due care to protect others from an unreasonable risk of harm. Science teachers and their supervisors have three basic duties. Failure to perform any of these could result in a legal finding that a teacher or a school administrator (or both) is liable for damages and a judgment and award against that teacher or school administrator (Council of Chief State Science Supervisors, no date, p. 2):

1. The duty of instruction. Teachers must instruct students prior to any laboratory activity, providing accurate, appropriate information about foreseeable dangers; identifying and clarifying any specific risks; explaining proper procedures/techniques; and describing appropriate behavior in the lab. These instructions must follow professional and district guidelines.

2. The duty of supervision. This includes not tolerating misbehavior, providing greater supervision in more dangerous situations, providing greater supervision to younger students and those with special needs, and never leaving students unattended.

3. The duty of maintenance. This requires that the teacher never use defective equipment, file written reports for maintenance or correction of hazardous conditions or defective equipment, establish regular inspections of safety equipment and procedures, and follow all guidelines for handling and disposing of chemicals.

Standards of Care for Student Safety

The courts have established that negligence may occur if teachers or school administrators' conduct falls below a standard of care established by law or profession. Standards of care are established not only by law and regulation but are also incorporated in building codes and guidelines established by voluntary associations.

In the event of student accident or injury, courts may consider whether the size of the laboratory facility and the number of students using the facility met standards of care. State laws and regulations governing class size are based on occupancy standards established by the Building Officials and Code Administrators International, Inc., and the National Fire Protection Association, Inc. (Roy, 1999). Both of these sets of standards call for 50 square feet of space per person in school laboratories or workshops. The National Science Teachers Association (NSTA) calls for a minimum of 45 square feet per student for a standalone laboratory and 60 square feet per student for a combination laboratory-classroom (Biehle et al., 1999, p. 55). This translates into at least 1,250 square feet for a laboratory and 1,440 square feet for a combined laboratory classroom. The NSTA recommends a maximum class size of 24 students in high school laboratory science classes.

The U.S. Occupational Safety and Health Administration (OSHA) establishes standards of care to protect the health and safety of all employees, including teachers and other school employees. One of the most important OSHA standards of care for school laboratories is the Laboratory Standard (29 CFR 1910.1450). This standard requires school science teachers to create and maintain a chemical hygiene plan (CHP). In most schools, a science teacher or teachers develop the CHP, which outlines policies, procedures, and responsibilities to increase student, teacher, and staff awareness of potentially harmful chemicals. The CHP requires proper labeling of all chemicals, using a Material Safety Data Sheet, which outlines important safety information, and safe storage. These data sheets must be made available to school employees and must be kept in a safe but easily accessible location. The National Institute for Occupational Safety and Health provides guides for proper separation of incompatible chemical families. Other OSHA standards governing laboratory safety include CFR, Part 29, 1910 (General Workplace Standards), 1910 Subpart Z (Exposure Standards), 1910.133 (Eyewear Standards), and 1910.1450 (Occupational Exposure to Hazardous Chemicals in Laboratories).

The U.S. Environmental Protection Agency (EPA) administers several laws and regulations affecting safety in high school science laboratories. These include (1) the Resource Conservation and Recovery Act, (2) the Emergency Planning and Right-to-Know laws and regulations, and (3) the Toxic Substances Control Act. To carry out provisions of the Resource Conservation and Recovery Act, EPA issues regulations and guidelines governing safe storage of laboratory chemicals, equipment, and supplies. Title III of this act governs emergency planning and right-to-know (about potentially hazardous chemicals), and Title IV governs chemical disposal. In implementing the Toxic Substances Control Act, EPA issues regulations and guidelines to protect indoor air quality. EPA provides a checklist for teachers to assess and improve indoor air quality, including items related specifically to school science laboratories (http://www.epa.gov/iaq/schools/tfs/teacher.html).

In addition to these federal regulations and guidelines, the American National Standards Institute (ANSI) has established voluntary standards for laboratory safety that include:

- ANSI Z358.1—guidelines for establishing the correct design, installation, use, and performance of emergency safety equipment.
- ANSI Z87—guidelines for protective equipment at easily accessible locations.

To help teachers and schools meet the growing body of standards of care, several organizations, ranging from the Council of Chief State Science Supervisors (no date) to Flinn Scientific, have created safety checklists. Many are readily accessible on the Internet (see Box 6-2). One company has developed comprehensive state-level guides available on CD-ROM, incorporating state regulations and guidelines, as well as federal and professional requirements (Jakel, Inc., 2005).

Current Patterns in Implementing Safety Policies

Although states, school districts, and professional associations make some efforts to alert schools and teachers about safety policies and practices, some evidence suggests that schools tend to react to accidents rather than taking positive action to avoid them. The costs of adequate safety are large. For example, between 2000 and 2003, the Chicago Public Schools spent $570,000 to conduct chemical sweeps in schools, at a cost of approximately $2,600 per school. When the Chicago science supervisor proposed a more serious and sustained investment in safety—including $3.3 million for initial equipment, teacher training, and policies for laboratory safety, followed by an annual investment of $1 million to continue inventories of chemicals, train teacher and supervisors, and employ safety specialists, the budget proposal was turned down (see Table 6-4).

While preventive safety measures are expensive, the costs of accidents and injuries may be even larger. Press reports indicate that some school and district officials do not make safety improvements until an accident occurs.

BOX 6-2   Laboratory Science Safety Checklists

> There are many sources of general safety checklists and action plans for teachers and school administrators concerned about laboratory safety. They include
>
> Council of Chief State Science Supervisors
>   (http://www.csss.enc.org/safety.htm)
> National Science Teachers Association
>   (http://www.nsta.org/positionstatementandpsid=32)
> National Science Education Leadership Association
>   (http://www.nsela.org/safesci17.htm)
> Flinn Scientific (a vendor of laboratory equipment and supplies)
>   (http://www.flinnsci.com/Sections/Safety/generalSafety/steps Prove.asp)
> Laboratory Safety Institute
>   (http://www.labsafety.org)

For example, in 2000, eight chemistry students in a Battle Creek, Michigan, school were severely burned when a teacher poured methanol into metal chloride salt. A ball of fire flashed across the teacher's desk and engulfed students sitting across from him. The teacher did not use a fume hood, because the one in his classroom forced observers to peer over his shoulder, preventing all students from watching. Following the accident, the district completed a previously planned renovation, providing every laboratory with a new fume hood that offers a better view of demonstrations (Hoff, 2003).

More recently, three students were burned at Federal Way High School near Seattle, Washington, when the teacher did a similar demonstration without a shield. A school spokeswoman commented, "None of our classrooms are set up that way" (Hagey, 2004). Since the accident, a state inspector from the state Department of Labor and Industries found five serious hazards in violation of state regulations, including: (1) emergency showers were not tested annually and emergency eyewashes were not tested weekly; (2) a district-wide chemical hygiene plan had not been implemented; (3) fume hoods were not tested to determine if they met national standards; (4) several bottles of acids and bases were stored on the floor of a fume hood, obstructing air flow and creating the risk of inhaling dangerous fumes; and (5) air sampling for formaldehyde exposure had not been carried out in biology labs (Maynard, 2004).

TABLE 6-4 Estimated Costs of Improving Laboratory Safety in Chicago Public Schools, 2004

Recommendation	Initial Cost	Annual Cost
1. Identify and codify laboratory safety procedures.	$20,000	$2,500
2. Establish clear accountability systems for the maintenance and management of chemical hygiene at local schools.	$100,000	$100,000
3. Establish a science safety manager position.	$80,000	$80,000
4. Identify one chemical hygiene specialist in each school.	$325,000	$325,000
5. Conduct priority removal of potentially hazardous chemicals that may remain in schools.	$400,000	$0
6. Deploy a system-wide web-based inventory system to collect and maintain an inventory of chemicals at each school.	$100,000	$50,000
7. Inventory existing chemical supply in schools as part of ongoing chemical hygiene plan. Remove hazardous chemicals from school science laboratories.	$265,000	$265,000
8. Provide baseline safety materials for all classrooms in which science laboratory investigations are taking place.	$1,800,000	$0
9. Roll out a four-tiered training plan focusing on laboratory safety.	$250,000	$250,000
10. Provide a set of "introduction to laboratory safety" lesson plans to be used by science teachers.	$10,000	$8,000
Total	$3,300,000	$1,000,000

SOURCE: Chicago Public Schools, Office of Math and Science.

Frequency of Accidents and Injuries

The weak and limited data available suggest that accidents are not uncommon in high school science laboratories. One study of injury claims related to school science in Iowa found that the number of claims rose from 674 in 1990-1993 to 1,002 in 1993-1996, and the cost to insurance companies rose from $1.68 to $2.3 million. The authors found that the number of lawsuits grew from 96 to 245, and awards in these suits grew from $566,305 to $1.2 million (Gerlovich et al., 2002).

Among teachers responding to a survey conducted in Texas in October 2000, 36 percent reported a total of 460 minor laboratory accidents during the 2000-2001 school year (Fuller et al., 2001, p. 9), and 13 percent reported a total of 85 major accidents requiring medical attention over the previous five years (Fuller et al., 2001, p. 10).

The lack of publicly available data on laboratory accidents and injuries may be due in part to the fact that many legal cases are settled before trial. As a result, there are few articles discussing legal precedents and findings in cases related to laboratory science (Standler, 1999).

Lack of Systemic Safety Enforcement

Over the past 10 years, several states have conducted surveys of laboratory safety in conjunction with teacher safety education workshops. States that have conducted surveys and workshops include Iowa (Gerlovich et al., 1998), Nebraska (Gerlovich and Woodland, 2000), North Carolina (Stallings, Gerlovich, and Parsa, no date), Wisconsin (Gerlovich, Whitsett, Lee, and Parsa, 2001), South Carolina (Sinclair, Gerlovich, and Parsa, 2003), and Alabama (Gerlovich, Adams, Davis, and Parsa, 2003). The results of these state surveys must be interpreted with caution, because responses were obtained from only small, self-selected samples of teachers, who may not be representative of the population of teachers more generally. For example, in Iowa, 617 surveys were mailed to participants who had agreed to attend safety training workshops, and 383 surveys were received at these workshops (Gerlovich et al., 2002). Surveys reflected the situation of at least one building in each of Iowa's area educational agencies, but it is not possible to determine whether the situation in other schools in those areas is the same or different.

Among the small group of teachers responding to the Iowa survey, nearly 70 percent worked in laboratories that were over 20 years old, making it less likely that they were in compliance with recent building codes. Less than 22 percent of the laboratories and 7 percent of the combined laboratory-classrooms included in this small sample complied with the NSTA standards calling for 45 square feet per student for laboratories and 60 square feet per student for combined laboratory-classrooms (Gerlovich et al., 2002). Most of the facilities included in the surveys had such basic safety features as ground fault interrupters on electrical outlets, ABC triclass fire extinguishers, and ANSI-approved eye protective equipment, but nearly 27 percent did not have a functioning eyewash station. About 37 percent of the teachers reported never receiving science safety training, and over 17 percent said they had received safety training more than 10 years earlier. Nearly 60 percent required students to sign safety contracts indicating they understood and agreed to follow safety procedures, and nearly 70 percent stored chemicals safely, based on chemical compatibility, rather than alphabetically. Further analysis of the survey data indicated that newer facilities (10 years old or less) generally had more square footage of floor space and were more likely to have two or more exits, compared with older facilities. The analysis also found that teachers who had received safety training within 10 years more frequently stored chemicals based on chemical compatibility than did teachers who had not been trained or had been trained more than 10 years previously.

Researchers in Texas distributed a safety survey to science teachers attending a conference in October 2000 and to teachers participating in 12 laboratory safety professional development sessions across the state (Fuller et al., 2001, p. 7). They received 590 responses. As in the Iowa study, the facilities in which most respondents taught were smaller than the size recommended by NSTA. Specifically, 94 percent of those who taught in laboratories indicated that these facilities were less than 1,200 square feet, indicating they did not meet the Texas recommendation of 50 square feet per student. Among respondents who worked in combined laboratory-classrooms, 70 percent reported room sizes of less than 1,000 feet, indicating that the rooms did not meet the requirement in Texas law of 50 square feet per student (Fuller et al., 2001, pp. 16-17). Many respondents also indicated that their schools did not follow standards of care regarding the availability and use of safety equipment, proper storage of chemicals, ventilation systems, and classroom communication (Fuller et al., 2001, p. 19). The Texas Hazard Communications Act requires all science teachers new to a school to participate in professional development activities focused on laboratory safety, but only 33 percent of respondents indicated that they had done so during the 2000-2001 school year.

Perhaps the most significant finding from the Texas survey was the positive and direct relationship between the number of students in a science class and the number of accidents. As student enrollments increased, so did the number of minor accidents. The authors recommended that school districts provide science laboratories of appropriate size (50 square feet per student) with appropriate storage space (15 square feet per student) and ventilation. They also recommended compliance with the recommended ratio of 25 students to 1 high school teacher (Fuller et al., 2001, p. 19).

Other data indicate that large class sizes may pose a threat to safety in school laboratories. Average science class sizes in California, 30.1 students per teacher in the 2003-2004 school year (California Department of Education, 2005), exceed the NSTA standard of 24 students per teacher in science classes conducting hands-on or inquiry activities. It may be extremely difficult for teachers in classes of 30 students to perform the "duty of supervision" and maintain safety during laboratory experiences. An earlier survey of Florida teachers published in 1988 indicated that they viewed the size of more than 55 percent of their classes to be "potentially unsafe" for laboratory work. The average class size viewed as "unsafe" was 31 students compared with an average class size of 23 students in the 45 percent of classes considered "safe" (Horton, 1988).

The data above suggest that schools and teachers need better training in safety. Although ad hoc workshops on laboratory safety can provide information that helps teachers and administrators enforce legal requirements for maintaining student safety, more sustained professional development may be required to create lasting changes in school safety, just as sustained professional development supports changes in teaching practices.

SUMMARY

Integrated laboratory-classrooms with flexible equipment and furnishing are ideal for supporting teaching and learning with laboratory experiences that are integrated into the flow of instruction. However, some schools are far from this ideal.

Direct observation and manipulation of many aspects of the material world require adequate laboratory facilities, including space for teacher demonstrations, student laboratory activities, student discussion, and safe storage space for supplies. Schools with higher concentrations of non-Asian minorities and schools with higher concentrations of poor students are less likely to have adequate laboratory facilities than other schools. In addition to lacking such adequate spaces for laboratory activities, schools with higher concentrations of poor or minority students and rural schools often have lower budgets for laboratory equipment and supplies than other schools. These disparities in facilities and supplies may contribute to the problem that students in schools with high concentrations of non-Asian minority students spend less time in laboratory instruction than students in other schools.

Laboratory safety is an area of growing concern in high school science, yet few systematic data are available on the current safety of facilities, equipment, and practices. School administrators and science teachers, who bear important responsibility for student safety, appear to receive little systematic safety training.

REFERENCES

Abramson, P. (2004, February). *9th annual school construction report: School planning and management.* Available at: http://www.peterli.com/global/pdfs/SPMConstruction2004.pdf [accessed Sept. 2004].

Agron, J. (2003, May). Growth spurt: Even as school districts and colleges continue to cut spending budgets, spending on construction booms. *American School and University Magazine.* Available at: http://www.asumag.com/mag/405asu21.pdf [accessed Sept. 2004].

Ainley, J. (1978). *An evaluation of the Australian science facilities program and its effect on science education in Australian schools.* Unpublished Ph.D. thesis, University of Melbourne.

Ainley, J. (1990). School laboratory work: Some issues of policy. In E. Hegarty-Hazel (Ed.), *The student laboratory and the science curriculum* (pp. 223-241). London, England: Routledge.

Arzi, H.J. (1998). Enhancing science education through laboratory environments: More than walls, benches and widgets. In B.J. Fraser and K.G. Tobin (Eds.), *International handbook of science education* (pp. 595-608). London, England: Kluwer Academic.

Banilower, E.R., Green, S., and Smith, P.S. (2004). *Analysis of data of the 2000 National Survey of Science and Mathematics Education for the Committee on High School Science Laboratories: Role and Vision* (September). Chapel Hill, NC: Horizon Research.

Biehle, J., Motz, L., and West, S. (1999). *NSTA guide to school science facilities.* Arlington, VA: National Science Teachers Association.

California Commission on Teacher Credentialing. (2004). *Science teacher preparation in California: Standards of quality and effectiveness for subject matter programs. A Handbook for teacher educators and program reviewers.* Sacramento: Author.

California Department of Education. (2005). *Ed-data: Profiles and reports.* Available at: http://www.ed-data.k12.ca.us/Navigation/fsTwoPanel.asp?bottom=%2Fprofile%2Easp%3Flevel%3D05%26reportNumber%3D16 [accessed Jan. 2005].

Council of Chief State Science Supervisors. (No date). *Science and safety: Making the connection.* Washington, DC: Author. Available at: http://csss.enc.org/safety.htm [accessed June 2005].

Englehardt, D.F. (1968). *Aspects of spatial influence on science teaching methods.* Unpublished Ed.D. thesis, Harvard University, School of Education.

Fuller, E.J., Picucci, A.C., Collins, J.W., and Swann, P. (2001). *An analysis of laboratory safety in Texas.* Austin: Charles A. Dana Center of the University of Texas at Austin.

Gerlovich, J.A., Adams, S., Davis, B., and Parsa, R. (2003). Alabama science safety: A 2001 status report. *ASTA News, 25*(1). Montgomery: Alabama State Teachers' Association.

Gerlovich, J.A., Parsa, R., Frana, B., Drew, V., and Stiner, T. (2002). Science safety status in Iowa schools. *Journal of the Iowa Academy of Science 109*(3, 4), 61-65.

Gerlovich, J.A., Whitsett, J., Lee, S., and Parsa, R. (2001). Surveying safety: How researchers addressed safety in science classrooms in Wisconsin. *Science Teacher, 68*(4), 31-35.

Gerlovich, J.A., Wilson, E., and Parsa, R. (1998). Safety issues and Iowa science teachers. *Journal of the Iowa Academy of Sciences, 105*(4), 152-157.

Gerlovich, J.A., and Woodland, J. (2000). Nebraska Secondary Science Teacher Safety Project: A 2000 status report. *Nebraska Science Teacher, 1*(1), 4-11.

Gohl, D. (2004). *Panel discussion of finances and resources.* Presentation to the Committee on High School Science Laboratories: Role and Vision, June 3-4, National Research Council, Washington, DC. Available at: http://www7.nationalacademies.org/bose/June_3-4_2004_High_School_Labs_Meeting_Agenda.html [accessed June 2005].

Hagey, J. (2004, January 31). Four hurt in lab explosion. *News Tribune* (Tacoma, WA).

Helderman, R. (2004, March 25). Institute plans VA biotech magnet school. *Washington Post.*

Henderson, A.T., and Mapp, K.L. (2002). *A new wave of evidence: The impact of school, family, and community connections on student achievement.* Austin: Southwest Educational Development Laboratory. Available at: http://www.sedl.org/connections/research-syntheses.html [accessed June 2005].

Henderson, D., Fisher, D.L., and Fraser, B.J. (2000). Interpersonal behaviour, laboratory learning environments, and student outcomes in senior biology classes. *Journal of Research in Science Teaching, 37,* 26-43.

Hoff, D.J. (2003). Science-lab safety upgraded after mishaps. *Education Week, 22*(33), 1, 20-21.

Horton, P. (1988). Class size and lab safety in Florida. *Florida Science Teachers Magazine,* spring.

Jakel, Inc. (2005). *The total science safety system.* Available at: http://showcase.netins.net/web/jakel/examples/examples.html [accessed June 2005].

LabPlan. (2004). *Introduction.* Available at: http://www.labplan.net/index.htm [accessed Oct. 2003].

Lazarowitz, R., and Tamir, P. (1994). Research on using laboratory instruction in science. In D. Gabel (Ed.), *Handbook of research on science teaching and learning* (pp. 94-128). New York: Macmillan.

Lidsky, A. (2004). *How financial and resource issues constrain or enable laboratory activities.* Presentation to the Committee on High School Science Laboratories: Role and Vision, June 3-4, National Research Council, Washington, DC. Available at: http://www7.nationalacademies.org/bose/June_3-4_2004_High_School_Labs_Meeting_Agenda.html [accessed June 2005].

Maynard, S. (2004, April 16). Hazards found in FWay schools science labs. *News Tribune.* Available at: http://www.labsafety.org/news/hazards_found_in_fway_schools_sc.htm [accessed June 2005].

Mulfinger, L. (2004). What is good science education, and whose job is it to support it? In D. Haase and S. Schulze (Eds.), *Proceedings of the Conference on K-12 Outreach from University Science Departments.* Raleigh: Science House, North Carolina State University.

National Center for Education Statistics. (2004a). *Contexts of elementary and secondary education: Trends in science and mathematics course taking.* Available at: http://nces.ed.gov/pubs2002/2002025_4.pdf [accessed Oct. 2005].

National Center for Education Statistics. (2004b). *The condition of education 2004. Societal support for learning.* Available at: http://www.nces.ed.gov/programs/coe/list/i6.asp [accessed July 2005].

National Research Council. (2003). *Bio 2010: Transforming undergraduate education for future research biologists.* Committee on Undergraduate Biology Education to Prepare Research Scientists for the 21st Century. Board on Life Sciences, Division on Earth and Life Studies. Washington, DC: The National Academies Press.

Novak, J.D. (1972). *Facilities for secondary school science teaching: Evolving patterns in facilities and programs.* Arlington, VA: National Science Teachers Association.

Roy, K. (1999). *For safety sake: One class size does not fit all!* Safe Science Series. Raleigh, NC: National Science Education Leadership Association. Available at: http://www.nsela.org/safesci7.htm [accessed June 2005].

Schenck, S., and Meeks, D. (1999). *Math and science programs: Making them count.* (ED 439-006.) New York: Office of the Comptroller, City of New York.

Schneider, M. (2002). *Public school facilities and teaching: Washington, DC and Chicago.* Paper commissioned by the 21st Century School Fund. Washington, DC: 21st Century School Fund. Available at: http://www.21csf.org/csf-home/publications.asp [Feb. 2005].

Schneider, M. (2004). *The educational adequacy of New Jersey public school facilities: Results from a survey of principals.* Newark: Education Law Center. Available at: http://www.edlawcenter.org/ELCPublic/elcnews_040510_jointpressrelease.htm [accessed May 2005].

Sheperd, R. (Ed.) (1974). Facilities for learning science [Special issue on the buildings and other facilities for the learning of science in Australian schools]. *Australian Science Teachers Journal, 20*(3), 4-76.

Sinclair, L., Gerlovich, J., and Parsa, R.A. (2003). South Carolina Statewide Science Safety Project. *Journal of the South Carolina Academy of Science, 1*(1), 19-27.

Smith, P.S., Banilower, E.R., McMahon, K.C., and Weiss, I.R. (2002). *The National Survey of Science and Mathematics Education: Trends from 1977 to 2000.* Chapel Hill, NC: Horizon Research. Available at: http://www.horizon-research.com/reports/2002/2000survey/trends.php [accessed May 2005].

Stallings, C., Gerlovich, J., and Parsa, R. (No date). *Science safety: A status report on North Carolina schools.* Raleigh: North Carolina Public Schools, Department of Public Instruction.

Standler, R. (1999). *Injuries in school/collage laboratories in the USA.* Available at: http://www.rbs2.com/labinj.htm [Oct. 2004].

Tamir, P. (1976). *The role of the laboratory in science teaching.* (Technical Report No. 10.) Iowa City: University of Iowa Science Education Center.

Trejos, N. (2003, August 25). Classroom's costly lessons: Cuts force teachers to forage for supplies. *Washington Post,* B1.

U.S. General Accounting Office. (1996). *School facilities: America's schools not designed or equipped for the 21st century.* (GAO-HEHS-95-95.) Washington, DC: Author. Available at: http://archive.gao.gov/t2pbat1/153956.pdf [accessed Oct. 2004].

Virginia Polytechnic Institute. (2004). *Chemistry outreach program.* Blacksburg, VA: Author. Available at: http://www.chem.vt.edu/chem-dept/mcl/index.html [accessed June 2005].

7

Laboratory Experiences for the 21st Century

Science education, of which laboratory experiences are a fundamental and unique part, is a critical component of education for the 21st century. Most policy makers and educators agree that scientific literacy is essential for all citizens in an increasingly technological world. At the same time, science education is essential to meeting the nation's needs for scientists and engineers in an era of growing global competition in research, development, and technological innovation. Yet in the United States, many people lack even a basic understanding of science. Because most Americans complete high school and the curriculum is designed to prepare young people both for employment and further study, high school science education has the potential to advance the dual goals of broad scientific literacy and preparation of the future technical and scientific workforce.

In this chapter we summarize the major findings and conclusions of the report and consider their implications for policy, practice, and research. We consider the role each conclusion plays in advancing a new vision of laboratory experiences in science education.

THE ROLE OF LABORATORY EXPERIENCES IN SCIENCE EDUCATION

The distinguishing feature of science is that explanations are required to correlate with observed data from nature. Scientists gather these data through direct observation, manipulation, and experimentation with natural phenomena. Because the subject matter of science is the material world, science education involves seeing, handling, and manipulating real objects and materials and teaching science involves acts of showing as well as of telling.

In the committee's view, science education includes learning about the methods and processes of scientific research (science process) and the knowledge derived through this process (science content). Science process centers on direct interactions with the natural world aimed at explaining natural phenomena. Science education would not be about science if it did not include opportunities for students to learn about both the process and content of science. Laboratory experiences, in the committee's definition, can potentially provide one such opportunity.

Most states and school districts continue to invest in laboratory facilities and equipment, many undergraduate institutions require completion of laboratory courses to qualify for admission, and some states require completion of science laboratory courses as a condition of high school graduation. These requirements exist without careful description of what is meant by a laboratory course. And, while some state and district policies appear to support laboratory experiences, others may hinder the design and implementation of effective laboratory learning experiences. The committee has identified science standards and assessments as two key policy drivers that shape the role of laboratory experiences in science education.

- State science standards that are interpreted as encouraging the teaching of extensive lists of science topics in a given grade may discourage teachers from spending the time needed for effective laboratory learning.
- Current large-scale assessments are not designed to accurately measure student attainment of the goals of laboratory experiences. Developing and implementing improved assessments to encourage effective laboratory teaching would require large investments of funds.

LABORATORY EXPERIENCES AND STUDENT LEARNING

The committee reviewed a wide body of research related to laboratory experiences and student learning. This review revealed a diffuse evidence base consisting of studies that vary widely in quality. The coherence of the body of evidence is complicated by a lack of clarity in the goals for laboratory experiences. As a first step to understanding the potential of laboratory experiences to advance science education, the committee defined laboratory experiences and identified seven goals.

- *Definition:* Laboratory experiences provide opportunities for students to interact directly with the material world (or with data drawn from the material world), using the tools, data collection techniques, models, and theories of science.

Goals of Laboratory Experiences

Laboratory experiences can help to enhance national scientific literacy and prepare the next generation of scientists and engineers by supporting students in attaining several educational goals:

- *Enhancing mastery of subject matter.*
- *Developing scientific reasoning.*
- *Understanding the complexity and ambiguity of empirical work.*
- *Developing practical skills.*
- *Understanding of the nature of science.*
- *Cultivating interest in science and interest in learning science.*
- *Developing teamwork abilities.*

Evidence on the Effectiveness of Laboratory Experiences

In reviewing the evidence on the effectiveness of laboratory experiences in helping students to attain these goals, the committee examined two somewhat distinct bodies of research. Each is designed to address a different question about the effectiveness of laboratory experiences.

Historically, laboratory experiences have been disconnected from the flow of science classes. Because this approach remains common today, we refer to these isolated interactions with natural phenomena as "typical" laboratory experiences. Research on typical laboratory experiences examines whether these encounters with the natural world, by themselves, contribute to students' science learning. Over the past 10 years, investigators have begun to develop a second body of studies that draw on principles of learning derived from cognitive psychology. This research has focused on development of instructional sequences that include laboratory experiences along with lectures, reading, and discussion. We refer to these instructional sequences including laboratory experiences as "integrated instructional units."

The earlier body of research, on typical laboratory experiences and the emerging research on integrated instructional units, yield different findings about the effectiveness of laboratory experiences in advancing the goals identified by the committee (see Table 7-1). Research on typical laboratory experiences is methodologically weak and fragmented, making it difficult to draw precise conclusions. The weight of the evidence from research focused on the goals of developing scientific reasoning and enhancing student interest in science showed slight improvements in both after students participated in typical laboratory experiences. Research focused on the goal of

Table 7-1 Attainment of Educational Goals in Different Types of Laboratory Experiences

Goal	Typical Laboratory Experiences	Integrated Instructional Units
Enhancing mastery of subject matter	No better or worse than other modes of instruction	Increased mastery compared to other modes of instruction
Developing scientific reasoning	Aids development of some aspects	Aids development of more sophisticated aspects
Understanding complexity and ambiguity of empirical work	Inadequate evidence	Inadequate evidence
Developing practical skills	Inadequate evidence	Inadequate evidence
Understanding of the nature of science	Little improvement	Some improvement when explicitly targeted at this goal
Cultivating interest in science	Some evidence of increased interest	Evidence of increased interest
Developing teamwork skills	Inadequate evidence	Inadequate evidence

student mastery of subject matter indicates that typical laboratory experiences are no more or less effective than other forms of science instruction (such as reading, lectures, or discussion).

A major limitation of the research on integrated instructional units is that most of the units have been used in small numbers of science classrooms. Only a few studies have addressed the challenge of implementing—and studying the effectiveness of—integrated instructional units on a wide scale. The studies conducted to date indicate that the laboratory experiences and other forms of instruction included in these units show greater effectiveness for these same three goals (compared with students who received more traditional forms of science instruction): improving students' mastery of subject matter, developing scientific reasoning, and cultivating interest in science and science learning. Integrated instructional units also appear to be effective in helping diverse groups of students progress toward these three learning goals. Due to a lack of available studies, the committee was unable to draw conclusions about the extent to which either typical laboratory experiences or integrated instructional units might advance the other goals identified at the beginning of this chapter—enhancing understanding of the complexity and ambiguity of empirical work, acquiring practical skills, and developing teamwork skills.

The committee considers the evidence emerging from research on integrated instructional units sufficient to conclude:

- Four principles of instructional design can help laboratory experiences achieve their intended learning goals if (1) they are designed with clear learning outcomes in mind, (2) they are thoughtfully sequenced into the flow of classroom science instruction, (3) they are designed to integrate learning of science content with learning about the processes of science, and (4) they incorporate ongoing student reflection and discussion.

These principles, combined with the seven goals, offer first steps toward a more coherent vision of laboratory experiences. They provide a framework for curriculum developers, administrators, and teachers to use in reconsidering how laboratory experiences can be successfully incorporated into science courses. The emerging research on the uses of technology to support laboratory experiences reveals a promising avenue for both research and practice, particularly for its potential to allow students access to otherwise inaccessible phenomena.

CURRENT HIGH SCHOOL LABORATORY EXPERIENCES

Analysis of current classroom practice shows that high school students' current laboratory experiences rarely follow the design principles we have identified. We conclude:

- The quality of current laboratory experiences is poor for most students.

Furthermore, access to any type of laboratory experience is unevenly distributed. Students in schools with higher concentrations of non-Asian minorities spend less time in laboratory instruction than students in schools with fewer non-Asian minorities. Students in more advanced science classes spend more time in laboratory instruction than students enrolled in regular classes. At the same time, most students, regardless of race or level of science class, participate in a limited range of laboratory experiences that are not based on the design principles derived from recent research in science learning.

READINESS OF TEACHERS AND SCHOOLS TO PROVIDE LABORATORY EXPERIENCES

One important factor contributing to the weakness of current laboratory experiences is a lack of preparation and ongoing support for high school science teachers. Effective high school laboratory teaching requires both deep conceptual and procedural knowledge of science disciplines and also deep knowledge of student learning and teaching strategies appropriate to those disciplines. However, current undergraduate education of future science teachers does not provide these types of knowledge. Undergraduate science departments rarely provide future science teachers with laboratory experiences that are designed on the basis of the learning principles identified in the research.

Once on the job, science teachers have few opportunities to improve their laboratory teaching. Most professional development opportunities for current science teachers are limited in quality, availability, and scope and place little emphasis on improving laboratory instruction. In addition, few high school teachers have access to science curricula that are designed on the basis of research, and some teachers struggle with inadequate laboratory space and supplies.

- Improving high school science teachers' capacity to lead laboratory experiences effectively is critical to advancing the educational goals of these experiences. This would require major changes in undergraduate science education, including provision of a range of effective laboratory experiences for future teachers and developing more comprehensive systems of support for teachers.
- The organization and structure of most high schools impedes teachers' and administrators' ongoing learning about science instruction and implementing quality laboratory experiences.

The design principles and goals offer a framework for reevaluating undergraduate science education for teachers, just as they can advance laboratory experiences in elementary and secondary schools settings. In addition, as state policy makers and district and school administrators begin to give more explicit and coherent attention to laboratory experiences, they can also supply the tools and support teachers need to provide high-quality laboratory experiences. For example, professional development might be designed with an explicit focus on laboratory experiences and tied to teachers' work in classrooms. Teachers could be given more time to plan and share ideas, and the time-intensive aspects of providing high-quality laboratory instruction could be recognized.

Finally, safety issues emerged as an important but neglected aspect of laboratory experiences. Greater attention to safety issues in research, policy, and practice is warranted.

TOWARD THE FUTURE: LABORATORY EXPERIENCES FOR THE 21ST CENTURY

Moving toward improvement of laboratory experiences for the 21st century is constrained by weaknesses in definitions and research. Historically, researchers studying laboratory experiences have not agreed on a precise definition of "laboratory." Even today, educators, policy makers, and researchers have differing views of the role and goals of high school laboratory experiences. This fragmentation in research, policy, and practice has slowed research, development, and demonstration of improved laboratory experiences.

- Researchers and educators do not agree on how to define high school science laboratories or on their purposes, hampering the accumulation of evidence that might guide improvement in laboratory education. Gaps in the research and in capturing the knowledge of expert science teachers make it difficult to reach precise conclusions on the best approaches to laboratory teaching and learning.

The need to more carefully define the role and goals of high school science laboratories and measure progress toward attainment of those goals is given greater urgency in view of the multiple pressures placed on schools and districts to increase the performance of a diverse student body. The challenge of meeting the needs of students in cost-effective ways places great pressure on schools to reevaluate the apparently more expensive features of education, such as high school science laboratories.

Although more recent research has illuminated the design principles to guide improvement in laboratory teaching and learning, studies of the possibilities and challenges associated with scaling up promising approaches are in the early stages. In addition, mechanisms for sharing the results of the research that is available—both within the research community and with the larger education community—are so weak that progress toward more effective laboratory learning experiences is impeded.

The committee envisions a future in which the role and value of high school science laboratory experiences are more completely understood. The state of the research knowledge base on laboratory experience is dismal but, even so, suggests that the laboratory experiences of most high school students are equally dismal. Improvements in current laboratory experiences can be made today using emerging knowledge. Documented disparities to access should be eliminated now.

Systematic accumulation of rigorous, relevant research results and best practices from the field will clarify the specific contributions of laboratory experiences to science education. Such a knowledge base must be integrated with an infrastructure that supports the dissemination and use of this knowledge to achieve coherent policy and practice.

The committee suggests that partnerships may be most successful in addressing the weaknesses in current laboratory experiences and other problems we have outlined. Specifically, teachers, scientists, cognitive psychologists, education researchers and school systems, working together, are best able to design and test innovative approaches to laboratory experiences. Partnerships like these are well suited to the challenge of answering the many remaining questions about laboratory teaching and learning:

1. *Assessment of student learning in laboratory experiences*—What are the specific learning outcomes of laboratory experiences and what are the best methods for measuring these outcomes, both in the classroom and in large-scale assessments?

2. *Effective teaching and learning in laboratory experiences*—What forms of laboratory experiences are most effective for advancing the desired learning outcomes of laboratory experiences? What kinds of curriculum can support teachers and students in progress toward these learning outcomes?

3. *Diverse populations of learners*—What are the teaching and learning processes by which laboratory experiences contribute to particular learning outcomes for diverse learners and different populations of students?

4. *School organization for effective laboratory teaching*—What organizational arrangements (state and district policy, funding priorities and allocation of resources, professional development, textbooks, emerging technologies, and school and district leadership) support high-quality laboratory experiences most efficiently and effectively? What are the most effective ways to bring those organizational arrangements to scale?

5. *Continuing learning about laboratory experiences*—How can teachers and administrators learn to design and implement effective instructional sequences that integrate laboratory experiences for diverse students? What types of professional development are most effective to help administrators and teachers achieve this goal? How should laboratory professional development be sequenced within a teacher's career (from preservice to expert teacher)?

Improving the quality of laboratory experiences available to U.S. high school students in order to advance the educational goals identified in this report will require focused and sustained attention. By applying principles of instructional design derived from ongoing research, science educators can begin to more effectively integrate laboratory experiences into the science curriculum. The definition, goals, design principles, and findings of this report offer an organizing framework to begin the difficult work of designing laboratory experiences for the 21st century.

Appendixes

APPENDIX A

Agendas of Fact-Finding Meetings

FIRST FACT-FINDING MEETING
March 29-30, 2004

Monday, March 29
Open Session

10:10 a.m.	**Welcome**
	Martin Orland, director, Center for Education, National Research Council (NRC)
	Jean Moon, director, Board on Science Education, NRC
10:20 a.m.	**Discussion of the Charge with the Sponsor**
	Janice Earle, senior program director, Division of Elementary, Secondary, and Informal Education, National Science Foundation (NSF)
	James Lightbourne, senior advisor, NSF Directorate for Education and Human Resources
10:45 a.m.	**Discovery Learning and Discovery Teaching**
	David Hammer, professor, University of Maryland

11:45 a.m.	Lunch
12:45 p.m.	**High School Science Laboratories: Data from the 2000 National Survey of Science and Mathematics Education**

                  Sean Smith, senior research associate, Horizon Research, Inc.

1:30 p.m.	**Panel Discussion with National Leaders in Science Education**

                  Gerald F. Wheeler, executive director, National Science Teachers Association
                  Warren W. Hein, associate executive officer, American Association of Physics Teachers
                  Angela Powers, senior education associate, teacher training, American Chemical Society
                  Michael J. Smith, education director, American Geological Institute

The panelists will address the following topics:
1. The current role of labs in high school science education;
2. Resources the association provides to assist teachers with labs; and
3. A vision for the future of high school labs.

2:45 p.m.	Break
3:00 p.m.	**History of High School Science Curriculum Development**

                  Janet Carlson-Powell, associate director, Biological Sciences Curriculum Study

4:00 p.m.	Open Session Adjourns

Tuesday, March 30

10:15 a.m.	**History of NSF Programs to Improve High School Science**

                  Gerhard Salinger, program director, Division of Elementary, Secondary, and Informal Education, NSF

10:45 a.m.	**Technology and High School Science**
	Robert Tinker, president, The Concord Consortium
11:45 a.m.	**Lunch**
1:00 p.m.	**Project 2061 Evaluation of Science Texts and Supporting Materials**
	Jo Ellen Roseman, director, American Association for the Advancement of Science Project 2061 **George DeBoer**, deputy director, AAAS Project 2061
2:00 p.m.	**Open Session Adjourns**

SECOND FACT-FINDING MEETING
June 3-4, 2004

Thursday, June 3
Open Session

9:00 a.m.	**Welcome**
	Jean Moon, director, Board on Science Education **Susan Singer**, chair, Committee on High School Labs
9:15 a.m.	**The Nature of Science and Scientific Research: Implications for High School Science Laboratories**
	Jane Maienschein, Arizona State University
	Questions speaker will address: (1) Briefly sketch the points of agreement and disagreement in current thinking about the nature of science (NOS) and how scientists work. (2) How can these views of NOS and how scientists work inform the goals and design of science education? (3) More specifically, how should/can this understanding inform the design of high school science lab experiences and their role in science education?

9:30 a.m.   **Discussion of Presentation**

10:00 a.m.   **Break**

10:10 a.m.   **Definition of Labs and Their Role in Science Education**

Robin Millar, University of York (UK)

Questions speaker will address:
(1) Outline your definition of laboratory work and the analytic framework that drives this definition. Under this definition, what distinguishes laboratory work from other aspects of instruction in science? How does inquiry fit into your definition?
(2) Given your definition of laboratory work, what unique role does the laboratory play in supporting students' learning in science? Or, put another way, what would be the consequences for students' learning in science if laboratory experiences, as you have defined them, were eliminated?
(3) What factors must be considered in determining the effectiveness of laboratory experiences? To what extent is it possible to derive a common set of characteristics of lab experiences that can be considered "good" or "effective" across a range of different learning goals and content areas?
(4) What are the most effective or most useful assessments of student learning in laboratory contexts?

10:30 a.m.   **Discussion of Presentation**

11:00 a.m.   **How Financial and Resource Issues Constrain or Enable Laboratory Activities**

James Guthrie, Vanderbilt University
Arthur Lidsky, Dober, Lidsky, Craig and Associates

Questions speakers will address:
(1) How do finances and other resources (including the costs of teacher training, space, equipment, technician support for teachers) enable or constrain

high school laboratory space, equipment, and activities?
(2) What is the range of lab experiences that schools with different levels of resources (including financial resources, lab space, lab equipment, and technology) provide?
(3) How should the physical and/or virtual laboratory be designed? What should it look like?

11:30 a.m. **Panel Discussion of Finances and Resources (each panelist will give 5 min opening comments)**

Daniel Gohl, principal, McKinley Technical High School, DC Public Schools
Shelley Lee, science education consultant, Wisconsin Department of Public Instruction
Lynda Beck, former assistant head of school, Phillips Exeter Academy
Kim Lee, science curriculum supervisor, Montgomery County Public Schools, VA

12:00 p.m. **General Discussion** (questions from committee and audience)

12:30 p.m. **Lunch**

1:30 p.m. **Organization and Administration of Schools to Sustain Instructional Improvement**

Adam Gamoran, University of Wisconsin

Questions speaker will address:
(1) What factors in the organization and administration of high schools and in education more generally enable sustained improvement in science instruction (including laboratory experiences)?
(2) What kinds of changes might be needed in the organization and administration of high schools to enhance the effectiveness of science labs?

1:50 p.m. **Discussion of Presentation**

2:20 pm. **Panel Discussion of School Leadership to Support Laboratory Experiences** (each panelist will give 5 min opening comments)

Daniel Gohl, McKinley Technical High School, DC Public Schools
Shelley Lee, science education consultant, Wisconsin Department of Education
Kim Lee, science curriculum supervisor, Montgomery County Public Schools, VA

Questions panelists will address:
(1) What sort of leadership is needed from the science department, from the school principal, and from the state to support sustained improvement in laboratory instruction?
(2) How can science teachers, state and local administrators, and outside organizations (e.g., scientists) develop relationships that enable and sustain quality laboratory instruction?

2:50 p.m. **Questions from Committee and Audience**

3:15 p.m. **Open Session Adjourns**

Friday, June 4
Open Session

9:00 a.m. **How Students Learn Science in Different Forms of Laboratory Experience: Focus on Technology**

Marcia Linn, University of California Berkeley

Questions speaker will address:
(1) What do we know about the role of technology (in all of its forms) in science learning?
(2) What are the unique contributions that technology can make to science learning?
(3) What does the evidence about students' learning and technology imply for developing a vision for the role of high school laboratories in science education?

9:20 a.m. **Discussion of Presentation**

10:00 a.m.	**Break**
10:15 a.m.	**How Teachers Learn and Work—Specifically Related to Labs**

Kefyn Catley, Vanderbilt University
Mark Windschitl, University of Washington

Questions speakers will address:
(1) What knowledge and skills are required to successfully design and carry out different forms of laboratory experiences?
(2) To what extent do teachers' current preparation and professional development provide them with these knowledge and skills?

11:00 a.m.	**Questions from Committee and Audience**
11:30 a.m.	**Lunch**
12:30 p.m.	**State Science Standards and Laboratory Assessment in New York: A Case Study**

Audrey Champagne, SUNY Buffalo
Thomas Shiland, Saratoga Springs High School

1:00 p.m.	**Science Standards and Assessment Across the 50 States**

Arthur Halbrook, Council of Chief State School Officers

1:30 p.m.	**Discussion of Presentations**

Questions speakers will address:
(1) What are the challenges in assessing students' learning from laboratory experiences?
(2) What is the current state of state standards and assessment practices and how do they constrain or enable what can be done in labs?

2:15 p.m.	**Open Session Adjourns**

THIRD FACT-FINDING MEETING
July 12-13, 2004

Monday, July 12
Open Session

9:30 a.m.   Coffee Break

9:40 a.m.   How Students Learn Science: The Role of Laboratories

Philip Bell, University of Washington
Richard Duschl, Rutgers University
Norman Lederman, Illinois Institute of Technology

Questions speakers will address:
(1) What does the research evidence suggest is the unique contribution of labs to students' learning in science? Put another way, what would be the consequences for students' learning in science if laboratory experiences were eliminated? Include in your response some discussion of content versus process goals and the extent to which these can be considered separately.
(2) What key principles can be drawn from our knowledge of students' learning in labs and science learning in general to guide both the design of future laboratory experiences and how they are integrated into the overall flow of science instruction? To what extent are these design principles shaped by which broad goals for science education are considered highest priority (for example, motivating students to continue learning science vs. training future scientists vs. developing science literacy for all)?

10:40 a.m.   Discussion of Presentations

11:30 a.m.   Lunch

12:30 p.m.   How Students Learn Science: Diverse Learners and Labs

Okhee Lee, University of Miami (by speakerphone)
Sharon Lynch, George Washington University
Kenneth Tobin, City University of New York

1:45 p.m.   **Discussion of Presentations**

Questions speakers will address:
(1) Do students of varying backgrounds (SES, ethnicity, language, disability, gender) have equal access to laboratory experiences? If not, what are the factors that lead to unequal access and what are the consequences for students' learning in science and the pathways in education and employment available to them?
(2) Do students of varying backgrounds learn science more effectively through laboratory experiences?
(3) Do labs motivate students of varying backgrounds to continue science education? If so, is this because labs help students see themselves as part of a "community of learners" in scientific discovery?
(4) How should future laboratory experiences be designed and delivered in order to reach students of varying backgrounds?

2:30 p.m.   **Students' Pathways in Science: Labs and Workforce Skills**

Samuel Stringfield, Johns Hopkins University

3:15 p.m.   **Break**

3:30 p.m.   **Students' Pathways: Labs for Biotechnology Careers**

Ellyn Daugherty, San Mateo High School, San Mateo, CA
Elaine Johnson, San Francisco Community College and Bio-Link

Questions speakers will address:
(1) What pathways in education and employment do high school science students follow?
(2) What is the role of technical education and the business community in enhancing the effectiveness of high school labs?
(3) What is the role of laboratory experiences in helping students pursue alternative pathways in education and employment?

4:20 p.m.   **Discussion of Presentations**

5:00 p.m.   **Adjourn for the Day**

Tuesday, July 13
Open Session

8:30 a.m.	**Students' Pathways: High School Labs and College Achievement**

Philip Sadler, Harvard University
Robert Tai, University of Virginia

9:00 a.m.   **Discussion of Presentation**

9:45 a.m.   **How Master Teachers Design and Carry Out Laboratory Experiences**

Nina Hike-Teague, Curie High School, Chicago, IL
Gertrude Kerr, Howard High School, Howard County, MD
Margot Murphy, George's Valley High School, ME
Phil Sumida, Maine West High School, Des Plaines, IL
Robert Willis, Ballou High School, Washington, DC

Questions panelists will address:
(1) Why and how do you incorporate laboratory experiences into instruction? How would you describe your specific learning goals for students in labs or lab-like situations?
(2) What are the biggest challenges to incorporating lab experiences into your instruction and could you provide a short example of how you deal with what you see as the most critical challenge?

11:15 a.m.   **Open Session Adjourns**

APPENDIX B

Biographical Sketches of Committee Members and Staff

Susan Singer (*Chair*) is professor of biology at Carleton College, where she has been since 1986. From 2000 to 2003 she directed the Perlman Center for Learning and Teaching, then took a research leave supported by a Mellon new directions fellowship. She chaired the Biology Department from 1995 to 1998 and was a National Science Foundation program officer for developmental mechanisms from 1999 to 2001. In her research, she investigates the evolution, genetics, and development of flowering in legumes; many of her undergraduate students participate in this research. She is actively engaged in efforts to improve undergraduate science education and received the Excellence in Teaching award from the American Society of Plant Biology in 2004. She helped to develop and teaches in Carleton's Triad Program, a first-term experience that brings students together to explore a thematic question across disciplinary boundaries. She is a member of the Project Kaleidoscope Leadership Initiative national steering committee and has organized PKAL summer institutes and workshops. At the National Research Council, she was a member of the Committee on Undergraduate Science Education and the Steering Committee on Criteria and Benchmarks for Increased Learning from Undergraduate STEM Instruction; currently she serves on the Board on Science Education. She has B.S., M.S., and Ph.D. degrees, all from Rensselaer Polytechnic Institute.

Hubert M. Dyasi is professor of science education at the City University of New York. He teaches undergraduates and graduates and works collaboratively with the New York State Education Department and school districts to develop their science education programs and implement inquiry-based science in classrooms. He has served as a specialist on science inquiry in the Harvard Smithsonian–Annenberg video program ("Looking at Learning . . . Again") and in the Annenberg-CPB's Professional Development Workshop Series. He is a contributing author to *Designing Professional Development for Teachers of Science and Mathematics* (2003); *Foundations: A Monograph for Professionals in Science' Mathematics, and Technology Education* (1999); and *Crossing Borders in Literacy and Science Instruction: Perspectives on Theory and Practice* (2004). At the National Research Council, he has been a member of the National Science Resource Center Advisory Board, the Committee on the Development of an Addendum to the National Science Education Standards, the Committee on Science Education K-12, and the Working Group on Science Teaching Standards. He is a fellow of the National Institute for Science Education. He has a Ph.D. in science education from the University of Illinois at Urbana-Champaign.

Arthur Eisenkraft is distinguished professor of science education at the University of Massachusetts, Boston, where he also directs the Center of Science and Math in Context. He recently left the Bedford, New York public school system, where he taught physics and was a science coordinator for 28 years. He is a past president of the National Science Teachers Association and has been involved with a number of its projects, creating and chairing many of the competitions sponsored by the association. He has been a columnist and advisory board member of the science and math student magazine *Quantum*. He is director of Active Physics, which is introducing physics instruction for the first time to all high school students. He is also directing another curriculum project, Active Chemistry. He holds a U.S. patent for an improved vision testing system using Fourier optics. At the National Research Council, he was a member of the curriculum working group that helped develop the National Science Education Standards, the Committee on Learning Research and Educational Practice, the Committee on Attracting Science and Mathematics Ph.D.s to K-12 Education, and the Committee on Assessing Technological Literacy. He is a fellow of the American Association for the Advancement of Science, a recipient of the Presidential Award for Excellence in Science Teaching (1986) and the Disney Science Teacher of the Year (1991). He has B.S. and M.A. degrees from the State University of New York at Stony Brook and a Ph.D. from New York University.

Margaret Hilton (*Study Director*) is a senior program officer at the Center for Education. She has written several National Research Council workshop reports and contributed to consensus studies on educational research, international labor standards, and the information technology workforce. In 2003, she was guest editor of a special issue of *Comparative Labor Law and Policy*. Prior to joining the National Academies in 1999, she was employed by the

National Skill Standards Board. Earlier, she was a project director at the congressional Office of Technology Assessment. She has a B.A. in geography (with high honors) from the University of Michigan and a master of regional planning degree from the University of North Carolina at Chapel Hill.

Pamela J. Hines is a senior editor at the international weekly journal *Science* and is responsible for selection and review of research manuscripts, as well as developing special issues, review articles, and perspectives on various topics. Her area of particular research interest is stem cell research, and she has expanded *Science*'s leadership role in highlighting developmental neurobiology, developmental biology, and plant sciences. She has conducted research on chromatin, gene control, and the mechanisms of DNA replication in eukaryotes during early development. She has served as editor-in-chief of the *AWIS Magazine* (Association for Women in Science), as a member of the communications committee for Oberlin College, and as co-principal investigator of a research project entitled Science Controversies: On line Partnerships for Education, which promotes science education through innovative uses of technology. Her current professional activities include serving on the editorial committee for the International Society for Stem Cell Research and the committee to develop representation of education issues at *Science*. She has an A.B. from Oberlin College, an M.S. from the University of Wisconsin, and a Ph.D. from the Johns Hopkins University.

Michael Lach is director of science for the Chicago Public Schools, overseeing science teaching and learning in the more than 600 schools that make up the nation's third largest school district. He began teaching high school biology and general science at Alceé Fortier Senior High School in New Orleans in 1990 as a charter member of Teach For America, the national teacher corps. He then joined the national office of Teach For America as director of program design, developing a portfolio-based alternative certification system that was adopted by several states. Returning to the science classroom in 1994 in the New York City Public Schools and then to Chicago's Lake View High School, he was named one of Radio Shack's Top 100 Technology Teachers, earned national board certification, and was named Illinois physics teacher of the year. He has served as an Albert Einstein distinguished educator fellow, advising Michigan Congressman Vernon Ehlers on science, technology, and education issues. He was lead curriculum developer for the Looking at the Environment materials developed at the Center for Learning Technologies in Urban Schools at Northwestern University. He has written extensively about science teaching and learning for such publications as *The Science Teacher*, *The American Biology Teacher*, and *Scientific American*. He has a B.S. in physics from Carleton College and M.S. degrees from Columbia University and Northeastern Illinois University.

David Licata teaches chemistry and advanced placement chemistry at Pacifica High School in Garden Grove, California. He chairs the science department and teaches general chemistry and chemistry for teacher education at Coastline Community College. From 1986 to 1990 he was the American Chemical Society's manager of precollege programs; in 2004 he chaired High School Chemistry Teachers' Day for the society's spring national meeting. He has been a member of the web-assisted tools for chemistry developers online group, sponsored by the Fund for the Improvement of Postsecondary Education. In the course of 22 years in the classroom, he has taught grades 6 through college and published numerous laboratory experiments and activities. He was the education workshop director for the Chevron Petroleum Research Corporation during the 1990s; there he planned a three-day conference for 35 teachers of grades 6 through 12 each year. Working with researchers, he has developed notes, experiments, and illustrations to highlight the interdisciplinary nature of scientific research and to give teachers the tools to relate this to their students. He has a B.S. in chemistry (1976) and an M.A. in administration (1981) from the University of California, Irvine.

Jean Moon (*Senior Program Officer*) is director of the Board on Science Education at the National Research Council. Previously she was a program officer and education adviser at the ExxonMobil Foundation, where she oversaw the conceptual development of the foundation's precollege and higher education portfolio. She has been a scholar-in-residence at the University of Uppsala in Sweden, where she worked with the natural science faculty on the development of competencies and assessment strategies to build communication skills in undergraduate science courses. She has consulted with numerous school districts on classroom assessment, institutional and role change in the context of education, and the development of case studies for use in teacher education contexts. Her current work focuses on the intersection of science, science education, and policy. She has a Ph.D. in Urban Education from the University of Wisconsin-Milwaukee.

Nancy Pelaez is associate professor of biological sciences specializing in physiology at California State University, Fullerton, where she advises science teachers in the master of arts in teaching science program. She is a former biology and chemistry teacher, with 10 years of experience in Bogota, Colombia, and 3 years in the Indianapolis Public Schools. She developed and directed the scope and sequence of K-12 science education at Colegio Los Nogales. She was the recipient of a Howard Hughes Medical Institute fellowship for a doctorate in vascular physiology, and she currently extends her biomedical research dissertation work to investigate discrepancies between student explanations of human blood circulation and those of the scientific community. She supports the peer review of exemplary instructional materials as associate editor of the Multimedia Educational Resource for Learning and Online Teaching and as an advisory board member of the American Association for the Advancement of Science's BiosciEdNet digital portal. She has a B.S. in biology from Newcomb College of Tulane University (1976), K-12 California single-subject teaching credentials in both life science and physical science from Mills College (1989), and a Ph.D. in physiology and biophysics from the Indiana University School of Medicine (1999).

William A. Sandoval is associate professor in the Psychological Studies in Education division of the Graduate School of Education and Information Studies at the University of California, Los Angeles. His research concerns students' epistemological beliefs about science, scientific argumentation, inquiry teaching practices, and the design of science learning environments. He began his work as a member of the BGuILE project at Northwestern University and is now principal investigator of the CENSEI project, exploring how to leverage live scientific data for middle school science. He has written and spoken internationally on the design of science learning environments, student and teacher learning in science, and design-based research methods. He serves on the editorial board of the *Journal of the Learning Sciences* and *Science Education* and reviews for a number of education research journals. He was cochair of the 2004 International Conference of the Learning Sciences and past chair of the special interest group in education in science and technology of the American Educational Research Association. He is a member of the American Educational Research Association, the National Association of Research in Science Teaching, the Association for Educational Communications and Technology, and the International Society of the Learning Sciences. He has a B.S. in computer science from the University of New Mexico and a Ph.D. in learning sciences from Northwestern University (1998).

Heidi Schweingruber (*Program Officer*) is on the staff of the Board on Science Education and is co-study director for its study of science learning in kindergarten through eighth grade. Prior to joining the National Research Council, she was a senior research associate at the Institute of Education Sciences in the U.S. Department of Education. In her role there she served as a program officer for the preschool curriculum evaluation program and for a grant program in mathematics education. She was also a liaison to the Department of Education's Mathematics and Science Initiative and an adviser to the Early Reading First program. Before moving into policy work, she was the director of research for the Rice University School Mathematics Project, an outreach program in K-12 mathematics education, and taught in the psychology and education departments. She is a developmental psychologist with substantial training in anthropology. She has a Ph.D. in psychology from the University of Michigan (1997).

James Spillane is professor of human development and social policy, professor of learning sciences, and a faculty fellow at the Institute for Policy Research, Northwestern University, where he teaches in both the learning sciences and human development and social policy graduate programs. He is director of the Multidisciplinary Program in Education Sciences. He is principal investigator of the Distributed Leadership Studies, which is undertaking an empirical investigation of the practice of school leadership in urban elementary schools that are working to improve mathematics, science, and literacy instruction. He is associate editor of *Educational Evaluation and Policy Analysis* and serves on the editorial board of numerous journals. He is author of *Standards Deviations: How Local Schools Misunderstand Policy* (2004) and *Distributed Leadership* (2005). Recent articles have been published in the *American Educational Research Journal, Cognition and Instruction, Educational Evaluation and Policy Analysis*, and *Education Researcher*, among others. A graduate of the National University of Ireland, he has a Ph.D. from Michigan State University (1993).

Carl Wieman is distinguished professor of physics at the University of Colorado and winner of the 2001 Nobel Prize in physics for studies of the Bose-Einstein condensate. He has been a member of the National Academy of Sciences since 1995. His research has involved the use of lasers and atoms to explore fundamental problems in physics. His group has carried out a variety of precise laser spectroscopy measurements, including accurate measurements of parity nonconservation in atoms and the discovery of the anapole moment. He has received numerous honors and awards in addition to the Nobel, including the Benjamin Franklin Medal in Physics, the Schawlow Prize for Laser Science, and the R.W. Wood Prize. In addition to his research activities, he has been involved in innovations in undergraduate physics education and has given many presentations to high school classes and general audiences. He directs the physics education technology project, which creates online interactive simulations for learning physics, and he has developed a popular physics course for nonscientists. Since 2000 he has worked on the National Task Force for Undergraduate Physics, which emphasizes improving undergraduate physics programs as a whole. At the National Research Council, he is chair of the Board on Science Education, and was a member of the Committee on Undergraduate Science Education. His contributions to both research and education have been recognized by the Richtmyer Memorial Lecture Award of the American Association of Physics Teachers and the first Distinguished Teaching Scholar Award of the director of the National Science Foundation. He has a Ph.D. from Stanford University (1977).

Index

A

AAAS. *See* American Association for the Advancement of Science
AAPT. *See* American Association of Physics Teachers
Access, to large databases, 32
Accidents, frequency of, 186–187
American Association for the Advancement of Science (AAAS), 20, 28, 54, 63, 83
American Association of Physics Teachers (AAPT), 158
American Chemical Society, 64–65, 67
American College Testing Service, 47
American Geological Institute, 65
American Institute of Biological Sciences, 64
American Institute of Physics, 65
American Physiological Society, 64
Arons, Arnold, 24

Assessment
 large-scale, 68
 of student learning in laboratory experiences, 10, 200
 in support of learning, informing integrated instructional units, 81
Assistance, expert, providing to schools and teachers, 155–156

B

Benchmarks for Science Literacy, 28
BGuILE science instructional unit, 94, 105
Biological Sciences Curriculum Study (BSCS), 23, 154
Brunner, Jerome, 26
BSCS. *See* Biological Sciences Curriculum Study

Budgeting for laboratory facilities, equipment, and supplies, 173–174
Building Officials and Code Administrators International, Inc., 183

C

California Department of Education, 30–31
California Institute of Technology, 155
Center for Embedded Networked Sensing (CENS), 106
Changing goals
    for the nature of science, 23
    for science education, 22–23, 28–29
Chemical Education Materials group, 23
Chemical hygiene plan (CHP), 183
Chemistry That Applies (CTA), scaling up, 82–83
CHP. *See* Chemical hygiene plan
City University of New York, 155
Clearly communicating purposes, 101
CLP. *See* Computer as Learning Partner
Community-centered environments, informing integrated instructional units, 81
Complex phenomena and ideas, structured interactions with, 105
Computer as Learning Partner (CLP), 84–85
Computer technologies and laboratory experiences, 103–106
    computer technologies designed to support learning, 103–105
    computer technologies designed to support science, 105–106
    scaffolded representations of natural phenomena, 103–104
    structured interactions with complex phenomena and ideas, 105
    structured simulations of inaccessible phenomena, 104–105
Conclusions
    regarding current high school laboratory experiences, 6
    regarding definitions and goals of high school science laboratories, 2
    regarding effectiveness of laboratory experiences, 6
    regarding laboratory facilities and school organization, 8
    regarding state standards and accountability systems, 9
    regarding teacher preparation for laboratory experiences, 7
Continued learning about laboratory experiences, 10, 200
Course-taking, disparities in laboratory experiences by variation in, 120–121
CTA. *See* Chemistry That Applies
Cultivating interest in science and interest in learning science, 77

Current debates, 30–31
Current high school laboratory experiences, 6–9, 197
    conclusions regarding, 6
    laboratory facilities and school organization, 7–8
    state standards and accountability systems, 8–9
    teacher preparation for laboratory experiences, 7
Current laboratory experiences, 116–137
    features of, 119–120
    quality of current laboratory experiences, 123–133
    quantity of laboratory instruction, 118–123
    summary, 133–134
    the unique nature of laboratory experiences, 117–118
Current patterns in implementing safety policies, 184–186
    estimated costs of improving laboratory safety, 186
    laboratory science safety checklists, 185
Current state of teacher knowledge, in preservice education, 145–148
    uneven qualifications of preservice science education, 147–148
    uneven qualifications of science teachers, 145–147
Curricula. *See also* New science curricula, developing; Post-Sputnik science curricula
    changing roles of, 29–30
    influence on science instruction, 7, 61–64

D

Databases, access to large, 32
Daugherty, Ellyn, 65
Design of effective laboratory experiences
    clearly communicated purposes, 101
    integrated learning of science concepts and processes, 102
    ongoing discussion and reflection, 102
    principles for, 101–102
    sequenced into the flow of information, 4, 102
Developing new science curricula, 23–26
    new approaches included in post-Sputnik science curricula, 25
Developing practical skills, 77, 92–93
    evidence from research on integrated instructional units, 93
    evidence from research on typical laboratory experiences, 92–93
Developing scientific reasoning, 76–77, 90–92
    evidence from research on integrated instructional units, 91–92
    evidence from research on typical laboratory experiences, 90–91
Developing teamwork abilities, 77
Dewey, John, 20–21
Diffusion, across a selectively permeable membrane, 125
Disabilities Education Act, 50
Discovery learning and inquiry, 26–27

Discussion, ongoing, 102
Disparities
    in laboratory equipment, 179–180
    in supplies, 180–182
Disparities in laboratory experiences, 120–123
    by ethnicity, 122–123
    and science course offerings, 121–122
    variation in course-taking, 120–121
Disparities in laboratory facilities, 177–179
    by proportion of minority students, 178
    by proportion of students eligible for free or reduced-price lunch, 179
Diverse populations of learners, 10, 200
Diversity increases, 48–51
    linguistic and ethnic diversity, 49
    in schools, 27
    special educational needs, 49–51
    in U.S. science education, 48–51

E

The education context, 42–74
    policies influencing high school laboratory experiences, 51–67
    recent trends in U.S. science education, 43–51
    summary, 67–68
Educational goals. *See* Goals for laboratory experiences
Effectiveness of laboratory experiences, 4–6, 10, 86–101, 200
    conclusions regarding, 6
    description of the literature review, 86–88
    developing practical skills, 92–93
    interest in science and interest in learning science, 95–98
    mastery of subject matter, 88–90
    overall effectiveness of laboratory experiences, 99–101
    teamwork, 98–99
    understanding the nature of science, 93–95
Emergency Planning and Right-to-Know laws and regulations, 183
Empirical work, understanding the complexity and ambiguity of, 77
Enrollment increases, 48–51
    linguistic and ethnic diversity, 49
    special educational needs, 49–51
    in U.S. science education, 48–51
EPA. *See* U.S. Environmental Protection Agency
Estimated costs, of improving laboratory safety, 186
Ethnicity, disparities in laboratory experiences by, 122–123
Evidence from research on integrated instructional units
    on developing practical skills, 93
    on interest in science and interest in learning science, 97–98

on mastery of subject matter, 89–90
on teamwork, 98–99
on understanding the nature of science, 94–95

Evidence from research on typical laboratory experiences
on developing practical skills, 92–93
on interest in science and interest in learning science, 95–96
on mastery of subject matter, 88–89
on teamwork, 98
on understanding the nature of science, 94

Evidence on the effectiveness of laboratory experiences, 195–197
attainment of educational goals in different types of laboratory experiences, 196
principles of instructional design, 197

Examples
of high school chemistry laboratory experiences, 130
of integrated instructional units, 82–85

Examples of professional development focused on laboratory teaching, 151–156
13-week science methodology course, 152–153
Biological Sciences Curriculum Study, 154
Laboratory Learning: An Inservice Institute, 152
professional development partnerships with the scientific community, 154–155
Project ICAN, 153
providing expert assistance to schools and teachers, 155–156

Expert assistance, providing to schools and teachers, 155–156

F

Facilities, equipment
and safety, 168–192
disparities in laboratory equipment, 179–180
disparities in laboratory facilities, 177–179
disparities in supplies, 180–182
laboratory safety, 182–189
providing, 168–192
summary, 189
and supplies, 168–192
budgeting for laboratory facilities, equipment, and supplies, 173–174
designing laboratory experiences and facilities when resources are scarce, 175–177
laboratories on wheels, 176
laboratory design and student learning, 169–173

Feedback, 81
Fermi National Accelerator Laboratory, 132
Fred Hutchinson Cancer Research Center, 154
Frequency of accidents and injuries, 186–187
Future perspectives, 199–201
assessment of student learning in laboratory experiences, 10, 200

continued learning about laboratory experiences, 10, 200
diverse populations of learners, 10, 200
effective teaching and learning in laboratory experiences, 10, 200
school organization for effective laboratory teaching, 10, 200

G

General pedagogical knowledge, 142–143
GenScope program, 104
Goals for laboratory experiences, 3–4, 76–78
cultivating interest in science and interest in learning science, 77
developing practical skills, 77
developing scientific reasoning, 76–77
developing teamwork abilities, 77
in different types of laboratory experiences, attainment of, 196
enhancing mastery of subject matter, 76
understanding the complexity and ambiguity of empirical work, 77
understanding the nature of science, 77

H

Hall, Edwin, 19
Harvard University, 19
HHMI. *See* Howard Hughes Medical Institute
High school science
role and vision of laboratory experiences in, 16
and undergraduate science achievement, in U.S. science education, 47–48
High school science laboratories
committee definition of laboratory experiences, 3
conclusions regarding, 2
definitions and goals of, 2–4
goal of laboratory experiences, 3–4
History of laboratory education, 18–30
1850-1950, 18–22
1950-1975, 22–27
1975 to present, 27–30
changing goals for science education, 22–23, 28–29
changing goals for the nature of science, 23
changing role of teachers and curriculum, 29–30
development of new science curricula, 23–26
discovery learning and inquiry, 26–27
diversity in schools, 27
How People Learn, 79
Howard Hughes Medical Institute (HHMI), 64, 66, 154, 175

I

ICAN. *See* Project ICAN
Inaccessible phenomena, structured simulations of, 104–105
Injuries, frequency of, 186–187
Instruction, teachers' duty of, 182

Instructional design, principles of, 6, 197
Instructional design principles, quality of current laboratory experiences compared with, 123–127
Integrated instructional units, 82–85, 196
    assessment to support learning, 81
    in community-centered environments, 81
    Computer as Learning Partner, 84–85
    design of, 81–82
    effectiveness of, 5
    in knowledge-centered environments, 80–81
    in learner-centered environments, 79
    principles of learning informing, 79–81
    scaling up Chemistry That Applies, 82–83
    for science concepts and processes, 102
    ThinkerTools, 84
Interactions
    with complex phenomena and ideas, structured, 105
    with data drawn from the real world, 3, 32
    with simulations, 31–32
Interest in science and interest in learning science, 95–98
    evidence from research on integrated instructional units, 97–98
    evidence from research on typical laboratory experiences, 95–96
    student perceptions of typical laboratory experiences, 96–97

International comparative test results, 46–47
International Technology Education Association, 172–173
Internet links, 32, 120
Introductory Physical Science, 23–24

K

Kilpatrick, William, 21
Knowledge-centered environments, informing integrated instructional units, 80–81
Knowledge Integration Environment project, 92
Knowledge of assessment, 143–144

L

Laboratories on wheels, 176
Laboratory design and student learning, 169–173
Laboratory experiences, 127–131
    approaches to learning physics using a pendulum, 128–129
    attainment of educational goals in different types of, 196
    committee definition of, 3
    continued learning about, 10, 200
    defining, 31–34, 37
    diffusion across a selectively permeable membrane, 125
    disparities in, 120–123
    examples of, 130
    overall effectiveness of, 99–101
    quality of current laboratory experiences compared with a range of, 127–131

role in science education, 193–194
science courses and, 118–120
what students do in, 132–133
Laboratory experiences and facilities, designing when resources are scarce, 175–177
Laboratory experiences and student learning, 75–115, 194–197
computer technologies and laboratory experiences, 103–106
defining, 194–195
effectiveness of laboratory experiences, 86–101
evidence on the effectiveness of laboratory experiences, 195–197
goals of laboratory experiences, 76–78, 195
principles for design of effective laboratory experiences, 101–102
recent developments in research and design of laboratory experiences, 78–85
summary, 106–108
Laboratory experiences for the 21st century, 193–201
current high school laboratory experiences, 197
readiness of teachers and schools to provide laboratory experiences, 198–199
role of laboratory experiences in science education, 193–194
toward the future, 199–201

Laboratory facilities and equipment
role of the scientific community in providing, 65–66
and school organization, 7–8
Laboratory-focused curriculum, role of the scientific community in providing, 65
Laboratory Learning: An Inservice Institute, 152
Laboratory safety, 182–189
checklists for, 185
current patterns in implementing safety policies, 184–186
estimated costs of improving, 186
frequency of accidents and injuries, 186–187
lack of systemic safety enforcement, 187–189
liability for student safety, 182
standards of care for student safety, 183–184
Laboratory Science Teacher Professional Development Program, 154
Laboratory teaching and learning, scheduling, 157–159
Learner-centered environments, informing integrated instructional units, 79
Learners, diverse populations of, 10, 200
Learning goals, need for focus on clear, 6, 123–124
Liability for student safety, 182
teachers' duty of instruction, 182
teachers' duty of maintenance, 182
teachers' duty of supervision, 182

Limitations of the research, 86–87
Linguistic and ethnic diversity, 49
The literature review, 86–88
    description of, 86–88
        limitations of the research, 86–87
        scope of the literature search, 87–88

M

Maintenance, teachers' duty of, 182
Mann, Charles, 20
Mastery of subject matter, 88–90
    evidence from research on integrated instructional units, 89–90
    evidence from research on typical laboratory experiences, 88–89
Material Safety Data Sheets, 183
Miller, Jon, 43
Minority students, disparities in laboratory facilities by proportion of, 178

N

NAEP. *See* National Assessment of Educational Progress
National Aeronautics and Space Administration (NASA), 64
National Assessment of Educational Progress (NAEP), 1, 44–46, 56, 119–120
National Center for Education Statistics, 146
    Schools and Staffing survey, 177
National Education Association, 19
National Education Longitudinal Study, 122
National Fire Protection Association, Inc., 183
National Human Genome Research Institute, 67
National Institute for Occupational Safety and Health, 183
National Institutes of Health (NIH), 67
National Research Council (NRC), 2, 14, 57, 79, 129, 146, 149, 171
National Science Achievement test results, 44–46
National Science Education Standards (NSES), 26, 28, 54–56, 59–60, 63
National Science Foundation (NSF), 2, 14, 22–24, 29–30, 43, 59, 65, 106, 119, 175
National Science Teachers Association (NSTA), 159, 172, 183, 187–188
"Negligence," 182–183
New approaches, included in post-Sputnik science curricula, 25
New science curricula, developing, 23–26
New York, hands-on performance assessment of laboratory learning experiences in, 58
New York State Regents exam, 20, 58, 174
NIH. *See* National Institutes of Health
No Child Left Behind Act, 54, 57
Noble Foundation, 67
Northeastern University, 155
NRC. *See* National Research Council
NSF. *See* National Science Foundation
NSTA. *See* National Science Teachers Association

O

Organisation for Economic Co-Operation and Development (OECD), Programme for International Student Assessment, 44, 46–47
OSHA. *See* U.S. Occupational Safety and Health Administration

P

Partnership for the Assessment of Standards-Based Science (PASS), 59
Pedagogical knowledge
    content, 141–142
    general, 142–143
Pendulum, approaches to learning physics using, 117, 126, 128–129
Performance assessment of laboratory learning, 58–59
    experiences in New York, 58
    experiences in Vermont, 59
Physical manipulation, of the real-world substances or systems, 31
Physical Science Study Committee (PSSC), 22–24
Piaget, Jean, 24
PISA. *See* Programme for International Student Assessment
Polanyi, Michael, 23
Policies influencing high school laboratory experiences, 51–67
    influence of curriculum on science instruction, 61–64
    role of the scientific community, 64–67
    science standards and assessments, 53–61
    state high school graduation requirements, 51–52
    state requirements for higher education admissions, 52–53
Post-Sputnik science curricula, 22
    new approaches included in, 25
Practical skills, developing, 77, 92–93
Preservice science education, uneven qualifications of, 147–148
Principles of learning informing integrated instructional units, 79–81
    assessment to support learning, 81
    community-centered environments, 81
    knowledge-centered environments, 80–81
    learner-centered environments, 79
Professional development, partnerships with the scientific community, 154–155
Professional development for laboratory teaching, 149–156
    examples of professional development focused on laboratory teaching, 151–156
    potential of professional development for improved laboratory teaching, 150–151
Programme for International Student Assessment (PISA), 44, 46–47

Project ICAN, 153
Project Physics, 24
Project SEED, 67
PSSC. *See* Physical Science Study Committee
Public understanding of science, in the United States, 43–44
Purposes, clearly communicating, 101

Q

Qualifications
    of preservice science education, uneven, 147–148
    of science teachers, uneven, 145–147
Quality of current laboratory experiences, 6, 123–133
    comparison with a range of laboratory experiences, 127–131
    comparison with instructional design principles, 123–127
    isolation from the flow of science instruction, 124–126
    lack of focus on clear learning goals, 123–124
    lack of reflection and discussion, 127
    little integration of science content and science process, 126–127
    what students do in laboratory experiences, 132–133
Quantity of laboratory instruction, 118–123
    disparities in laboratory experiences, 120–123
    science courses and laboratory experiences, 118–120

R

Readiness of teachers and schools to provide laboratory experiences, 198–199
Reflection and discussion
    lack of, 127
    ongoing, 102
Remote access to scientific instruments and observations, 32
Representations of natural phenomena, scaffolded, 103–104
Research, development, and implementation of effective laboratory experiences, 9–11, 79–85
    assessment of student learning in laboratory experiences, 10, 200
    continued learning about laboratory experiences, 10, 200
    design of integrated instructional units, 81–82
    diverse populations of learners, 10, 200
    effective teaching and learning in laboratory experiences, 10, 200
    examples of integrated instructional units, 82–85
    principles of learning informing integrated instructional units, 79–81
    recent developments in, 78–85
    school organization for effective laboratory teaching, 10, 200

RE-SEED. *See* Retirees Enhancing Science Education through Experiments and Demonstration
Resource Conservation and Recovery Act, 183–184
Retirees Enhancing Science Education through Experiments and Demonstration, 155

S

School organization, for effective laboratory teaching, 10, 200
Schools and Staffing Survey, 177
Schwab, Joseph, 23, 26
Science achievement in secondary school, 44–47
    results of international comparative tests, 46–47
    results of National Science Achievement Tests, 44–46
    in U.S. science education, 44–47
Science content knowledge, 140–141
    little integration with science process, 126–127
Science course offerings, 118–119
    disparities in laboratory experiences by, 121–122
Science courses and laboratory experiences, 118–120
    features of current laboratory experiences, 119–120
    science course-taking, 118–119
Science for All Americans, 26
Science instruction, isolation from the flow of, 124–126
Science Laboratory Environment Inventory (SLEI), 96–97
Science standards and assessments, 53–61
    hands-on performance assessment of laboratory learning, 58–59
    implementing state standards, 57, 60–61
    state science assessments and the goals of laboratories, 55–57
    state science standards and the goals of laboratories, 54–55
The scientific community, 64–67
    providing laboratory facilities and equipment, 65–66
    providing laboratory-focused curriculum, 65
    providing student internships, 66–67
Scientific issues, making informed decisions about, 1
Scientific reasoning, developing, 76–77, 90–92
Scientific societies, 64
Silliman, Benjamin, 19
Simulations of inaccessible phenomena, structured, 104–105
SLEI. *See* Science Laboratory Environment Inventory
Special education, need for, 49–51
Sputnik. *See* Post-Sputnik science curricula
State high school graduation requirements, 51–52
State requirements for higher education admissions, 52–53
State science assessments and the goals of laboratories, 55–57
State standards, implementation of, 57, 60–61

State standards and accountability systems, 8–9
  conclusions regarding, 9
  and the goals of laboratories, 54–55
Structured interactions, with complex phenomena and ideas, 105
Structured simulations, of inaccessible phenomena, 104–105
Student activities included among laboratory experiences, 31–32
  access to large databases, 32
  enabling by Internet links, 32
  interaction with data drawn from the real world, 32
  interaction with simulations, 31–32
  physical manipulation of the real-world substances or systems, 31
  remote access to scientific instruments and observations, 32
Student perceptions of typical laboratory experiences, and interest in science and interest in learning science, 96–97
Student safety, standards of care for, 183–184
Students
  carrying out laboratory investigations, 1
  diverse populations of, 10, 200
  eligible for free or reduced-price lunch, disparities in laboratory facilities by proportion of, 179
  role of the scientific community in providing internships for, 66–67
Subject matter, enhancing mastery of, 76
Supervision, teachers' duty of, 182
Support
  for laboratory teaching, 156–159
  scheduling laboratory teaching and learning, 157–159
  for teachers with professional development, 156–157
Systemic safety enforcement, lack of, 187–189

T

Teacher and school readiness for laboratory experiences, 138–167
  summary, 160
  supporting laboratory teaching, 156–159
Teacher knowledge for a range of laboratory experiences, 139–145
  general pedagogical knowledge, 142–143
  knowledge of assessment, 143–144
  pedagogical content knowledge, 141–142
  science content knowledge, 140–141
Teachers
  changing roles of, 29–30
  knowledge in action, 144–145
  preparation for laboratory experiences, 7
  uneven qualifications of, 145–147

Teachers' capacity to lead laboratory experiences, 139–156
  current state of teacher knowledge—preservice education, 145–148
  professional development for laboratory teaching, 149–156
Teachers' duties
  instruction, 182
  maintenance, 182
  supervision, 182
"Teaching for understanding," 157
Teamwork, 98–99
  evidence from research on integrated instructional units, 98–99
  evidence from research on typical laboratory experiences, 98
Teamwork abilities, developing, 77
ThinkerTools, 81, 84, 94, 97–98, 103–104
13-week science methodology course, 152–153
TIMSS. See Trends in International Mathematics and Science Study
Toxic Substances Control Act, 183–184
Trends in International Mathematics and Science Study (TIMSS), 44, 46, 62–63
Trends in U.S. science education, 43–51
  high school science and undergraduate science achievement, 47–48
  public understanding of science, 43–44
  rising enrollments and increasing diversity, 48–51
  science achievement in secondary school, 44–47
  toward the future, 199–201
Typology
  of school laboratory experiences, 36
  of scientists' activities, 35

U

Understanding the nature of science, 77, 93–95
  evidence from research on integrated instructional units, 94–95
  evidence from research on typical laboratory experiences, 94
U.S. Census Bureau, 177
U.S. Constitution, 27
U.S. Department of Energy, 64, 154
U.S. Environmental Protection Agency (EPA), 183
U.S. General Accounting Office (GAO), 169, 177
U.S. Geological Survey, 65
U.S. Occupational Safety and Health Administration (OSHA), 183

V

Vanderbilt University, 154
Variety in laboratory experiences, 33–34
Vermont, hands-on performance assessment of laboratory learning experiences in, 59
Virginia Polytechnic Institute (VPI), 65–66, 175

von Liebig, Justus, 19
VPI. *See* Virginia Polytechnic
 Institute

W

Waterman, Alan, 22

Z

Zacharias, Jerrold, 22, 24